9th Young Researcher Meeting 2018

AA006776

Journal of Physics: Conference Series
Volume 1226

Salerno, Italy
10 – 13 July 2018

ISBN: 978-1-5108-8946-0
ISSN: 1742-6588

Printed from e-media with permission by:

Curran Associates, Inc.
57 Morehouse Lane
Red Hook, NY 12571

Some format issues inherent in the e-media version may also appear in this print version.

Institute of Physics
Dirac House, Temple Back
Bristol BS1 6BE UK

Phone: 44 1 17 929 7481
Fax: 44 1 17 920 0979

techtracking@iop.org

Additional copies of this publication are available from:

Curran Associates, Inc.
57 Morehouse Lane
Red Hook, NY 12571 USA
Phone: 845-758-0400
Fax: 845-758-2633
Email: curran@proceedings.com
Web: www.proceedings.com

Table of contents

Volume 1226

9th Young Researcher Meeting

10–13 July 2018, Salerno, Italy

Accepted papers received: 8 April 2019
Published online: 10 June 2019

Preface

Papers

Applied Physics

Astrophysics and Cosmology

Detectors for particle physics

Geophysics

Theoretical Physics

IOP Conf. Series: Journal of Physics: Conf. Series **1226** (2019) 011001 doi:10.1088/1742-6596/1226/1/011001

Young Researcher Meeting, Salerno 2018

Fabio Agostini, Giulio Francesco Aldi, Claudia Antolini, Guerino Avallone, Giordano Cattani, Alessandro Cultrera, Domenico D'Agostino, Rocco D'Agostino, Ivan Delvecchio, Cinzia Di Giorgio, Marco Di Stefano, Paola Forlenza, Nicola Funicello, Luca Giovannelli, Laura Iemmo, Giuseppe Gaetano Luciano, Giuseppe Luongo, Simona Mancini, Eloisa Menegoni, Marina Migliaccio, Ferdinando Napolitano, Lorenzo Pagnanini, Fabrizio Piacentini, Davide Pietrobon, Emanuela Pusceddu, Cristina Ripoli, Matteo Serra, Francesco Stellato

IPN INTERNATIONAL PHYSICISTS NETWORK

YRM YOUNG RESEARCHER MEETING

Preface

The Young Researcher Meeting was founded in 2009 with the aim to gather undergraduate, graduate and PhD students together with young researchers to discuss the latest news, current progress and developments in various areas of physics. Held every year, the meeting offers the possibility to young physicists to present their original work and results, as well as to share their expertise and knowledge with the community. Another stimulating purpose of the meeting is to create an international network of researchers to promote collaborations and cross-fertilisation among different research areas.

The format chosen for the conference is an informal meeting where the attendees introduce their work with a brief presentation and/or a poster contribution.

The 9th edition of the Young Researcher Meeting was held at the University of Salerno, from 10th to 13th of July 2018 and it has been jointly organised by the Department of Physics "E. R. Caianiello" of the University of Salerno, the CNR-Spin and the International Physicists Network.

The event gathered 70+ participants affiliated with universities and research facilities in Italy and abroad. The plenary sessions covered several research areas, ranging from Astrophysics and Cosmology to Biophysics, Particle Physics, Condensed Matter, Theoretical Physics and Geophysics. The program also included a *lectio magistralis* by Prof. Antonio Di Bartolomeo from the University of Salerno on "The graphene-semiconductor Schottky junction" and three invited speakers: Dr Karoline Schaeffner from GSSI & INFN - LNGS, Dr Valerio Bozza from Salerno University & INFN and Dr Barbara Capone, Marie Curie Fellow at the University of Roma 3.

Moreover, the program included both a poster session during which the attendees had the possibility to taste local wine and chocolate, and a social dinner in a local restaurant with a breathtaking sea view of the Amalfi coast, a UNESCO World Heritage Site.

This year, participants were given the opportunity to take a tour of the Vesuvius National Park, guided by Prof. Roberto Scarpa, Director of the Physics Department at Salerno University and leading expert on Earthquake and Volcano physics.

The present volume of papers consists of a selection of the contributions presented at the conference as either talks or posters. These cover topics on Astrophysics and Cosmology, Biophysics, Theoretical and Particle Physics, and Condensed Matter.

IOP Conf. Series: Journal of Physics: Conf. Series **1226** (2019) 011001 doi:10.1088/1742-6596/1226/1/011001

ACKNOWLEDGEMENTS

The organisers of the 9th Young Researcher Meeting would like to thank all the participants for their scientific contributions and all the people who made this conference a very smooth and pleasant event.

They would like to thank the University of Salerno for hosting the conference; the Università degli Studi di Salerno, the Department of Physics "E. R. Caianiello" and Biofotonica for financial support; the CNR-Spin for its support and the INGV for the tour in the Parco Nazionale del Vesuvio.

They are grateful to Tenuta Cavaliere, Arte Terra, Consorzio provinciale Olivicoltori and Workingas for providing the local products offered to the participants during the Poster&Wine session. We also acknowledge the support of the companies Workingas and Biofotonica.

The organisers would also like to thank the Director of the Department of Physics of the University of Salerno, Prof. Roberto Scarpa, for his advice and guidance during the visit to the Parco Nazionale del Vesuvio, Prof. Antonio Di Bartolomeo for his enthusiasm in accepting to deliver the *lectio magistralis*, Dr Karoline Schaeffner, Dr Valerio Bozza and Dr Barbara Capone for accepting to participate as invited speakers.

Finally, the organisers would like to thank and acknowledge the work of all the fellow scientists who contributed reviewing the papers published in this volume, thus guaranteeing they met the expected scientific standards of the conference and the journal.

UNIVERSITÀ DEGLI STUDI DI SALERNO

Dipartimento di Fisica E.R. Caianiello

INGV

SPIN

BIOFOTONICA The Wizard Technology

WORKINGAS®

CONSORZIO PROVINCIALE OLIVICOLTORI SALERNO

ArteTerra Agriofficina L'eccellenza della Nocciola

TENUTA CAVALIER PEPE Azienda Vitivinicola Irpina in Campania

IOP Conf. Series: Journal of Physics: Conf. Series **1226** (2019) 011002 doi:10.1088/1742-6596/1226/1/011002

Peer review statement

All papers published in this volume of *Journal of Physics: Conference Series* have been peer reviewed through processes administered by the proceedings Editors. Reviews were conducted by expert referees to the professional and scientific standards expected of a proceedings journal published by IOP Publishing.

IOP Conf. Series: Journal of Physics: Conf. Series **1226** (2019) 012001 doi:10.1088/1742-6596/1226/1/012001

Effect of Phase Shift on Mixed Convection in a Rectangular Vented Cavity filled with a Nanofluid and Submitted to Periodic Heating

I. Arroub[1*], A. Bahlaoui[1], K. Ezzaraa[1], A. Raji[2], M. Hasnaoui[3] and M. Naïmi[2]

[1] Team of Applied Physics and New Technologies (EPANT), Polydisciplinary Faculty, Béni-Mellal, Morocco;
[2] Laboratory of Flows and Transfers Modeling (LAMET), Faculty of Sciences and Technics, Béni-Mellal, Morocco;
[3] Laboratory of Fluid Mechanics and Energetics (LMFE), Faculty of Sciences Semlalia, Marrakech, Morocco;

E-mail: ismail.arroub@gmail.com

Abstract. The characteristic of flow and heat transfer of Al_2O_3-water nanofluids flowing through a horizontal ventilated cavity is investigated numerically. The bottom wall is subjected to a sinusoidal hot temperature profile, whereas the other boundaries are assumed to be thermally insulated. In fact, the flow comes forcedly into the system from the bottom of the left vertical wall and leaves it from the top of the right one by suction. The simulations focus on the effects of different key parameters, such as Reynolds number ($200 \le Re \le 5000$), nanoparticle's concentration ($0 \le \phi \le 0.05$) and phase deviation of the sinusoidal heating (γ equal 0 or π), on the flow and thermal patterns and heat transfer performances. The obtained results show that the presence of nanoparticles increases the heat transfer and the mean temperature within the cavity. In addition, the phase shift of the heating temperature may lead to periodic solutions for weaker values of Re and contributes to an increase or a decrease of heat transfer depending on the value of ϕ and the convection regime.

1. Introduction

Combined free and forced convection heat transfer in a vented cavity is present in many practical engineering applications, such as solar energy systems, design of the heat exchangers, and cooling of electronic equipment. These systems have been widely used throughout the world. Thus few percent improvement on their performance could have significant impact on the energy saving. On the other hand, low thermal conductivity of conventional fluids such as water, oil, and ethylene glycol mixture that have been usually used in these systems is a serious limitation in improving the performance and compactness of these devices. So it would be very useful to improve the thermal characteristics of fluids applied in such devices. An innovative way of increasing the thermal conductivity of fluid is to disperse small solid particles into the base fluid; the mixture is called "nanofluid" [1].

Since higher heat transfer rates are required in modern industrial applications, mixed convection mechanism is often recommended in such situations, where natural or forced convection is not able to provide cooling effectiveness. Over the past recently years, mixed convection problem in nanofluids with varying heating temperature profile is mostly associated to lid-driven cavity. In that context, it conveys to mention the numerical study of Arani et al [2] in the case of a lid-driven square cavity filled with Cu-water nanofluid and exposed to sinusoidal heating on sidewalls. The obtained results indicate that for a constant Grashof number/(Reynolds number) the rate of heat transfer increases/(decreases) while decreasing the Richardson number. On the contrary, the volume fraction of nanoparticles affects positively such a quantity. The same problem was analyzed by Kefayati [3] for non-Newtonian aluminum-water nanofluids by using finite difference Lattice-Boltzmann method. It was observed that an augmentation of the power-law index or the Richardson number influences negatively the heat transfer enhancement by the nanoparticles. Other authors investigated mixed convection of nanofluids in a lid-driven shallow enclosure heated from one side. In this context, Goodarzi et al. [4] and Karimipour et al. [5] studied numerically mixed convection of Cu-water nanofluids, with cold and hot moving wall respectively. Their results showed that the average Nusselt number is augmented by increasing the volume fraction of nanoparticles.

Nanofluid mixed convection heat transfer occurs in other ways (case of ventilated cavities in semi-confined media). Over the last decades, few investigations on mixed convection of nanofluids in ventilated cavities with uniform heating (by imposing either temperature or heat flux on the heat source) have been realized. Such a problem is of great importance in various technological applications. Hence, in the case of applied uniform heat flux, Shahi et al. [6] conducted a numerical study of mixed convection in a partially heated square ventilated cavity, filled with copper-water nanofluid. The reported results show an increase of heat transfer rate at the heat source surface and a decrease in the mean bulk temperature with the solid concentration. The same problem was reconsidered by Mahmoudi et al. [7] for different locations of inlet and outlet ports in the case of total constant heating. It was found that the presence of nanoparticles is more effective in Bottom-Top configuration than in the other considered configurations, while the solid concentration increase has the least effects in Top-Top arrangement. In the case of a uniform imposed temperature, the problem of unsteady mixed convection flows through an Al_2O_3-water nanofluid in a square cavity with inlet and outlet ports due to incoming flow oscillation was performed numerically by Sourtiji et al. [8] for the Top-Bottom configuration. The results presented show an enhancement of heat transfer accompanying the addition of nanoparticles for all the Strouhal and Richardson numbers considered.

In electronic boards, sometimes there is an electronic component disposed along a plate and which dissipates heat while the temperature of its surface remains either uniform or varies periodically. The modeling of such a heating mode could be done by assimilating this component by a heating plate having a sinusoidally variable temperature. In this context, such problem was treated by the authors. Thus, the aim and scope of the current study is to simulate numerically mixed convection heat transfer characteristics of water-based Al_2O_3 nanofluid flowing through a rectangular cavity, with inlet and outlet ports. The buoyancy driven flow is due to imposed sinusoidal hot temperature profile on the lower horizontal wall of the cavity and the external incoming cold flow. The consequence of varying Reynolds number, nanoparticle concentration and phase deviation on the dynamical and thermal characteristics of the flow and heat transfer performances are investigated and discussed.

2. Problem definition and mathematical formulation

Figure 1 shows a schematic of the case considered in the present study. The problem deals with a two-dimensional vented enclosure of length L', height H' and having an aspect ratio A = 2, and filled with water-Al_2O_3 nanofluid. The enclosure is heated from the bottom wall by a sinusoidal hot temperature (Figure 2) whereas the other solid boundaries are assumed to be thermally insulated. The cavity is subjected to a sucked imposed external flow entering the cavity from the left-bottom opening and leaving from the right-upper one. These openings have a constant relative dimension, B = 1/4. Thus, it is assumed that both the fluid phase and nanoparticles are in thermal equilibrium and the shape and size of solid particles are assumed to be uniform. Also, the thermo-physical properties of the nanofluid are constant except for the density variation, which is approximated by the Boussinesq model.

Therefore, using the following dimensionless variables:

$$A = L'/H', \quad B = h'/H', \quad x = x'/H', \quad y = y'/H', \quad u = u'/u'_o, \quad v = v'/u'_o, \quad t = t'u'_o/H', \quad \Psi = \Psi'/u'_o H',$$
$$T = (T' - T'_C)/(T'_H - T'_C), \quad \Omega = \Omega' H'/u'_o, \quad Pr = \nu_f/\alpha_f, \quad Ra = g\beta_f(T'_H - T'_C)H'^3/\alpha_f \nu_f, \quad Re = u'_o H'/\nu_f, \quad Ri = Ra/Re^2 Pr$$

The mixed convection governing equations using the vorticity-stream function formulation (Ω-Ψ), can be written in the non-dimensional form as follows:

$$\frac{\partial \Omega}{\partial t}+u\frac{\partial \Omega}{\partial x}+v\frac{\partial \Omega}{\partial y}=\frac{Ra}{Re^2 Pr}\left[\left(\frac{\phi}{(1-\phi)\frac{\rho_f}{\rho_s}+\phi}\right)\frac{\beta_s}{\beta_f}+\frac{1}{\left(\frac{\phi}{(1-\phi)}\frac{\rho_s}{\rho_f}+1\right)}\right]\frac{\partial T}{\partial x}+\frac{1}{Re}\left[\frac{1}{(1-\phi)^{2.5}\left(\phi\frac{\rho_s}{\rho_f}+(1-\phi)\right)}\right]\left(\frac{\partial^2 \Omega}{\partial x^2}+\frac{\partial^2 \Omega}{\partial y^2}\right) \quad (1)$$

$$\frac{\partial T}{\partial t}+u\frac{\partial T}{\partial x}+v\frac{\partial T}{\partial y}=\frac{1}{Re\,Pr}\left(\frac{\frac{\lambda_{nf}}{\lambda_f}}{(1-\phi)+\phi\frac{(\rho c_p)_s}{(\rho c_p)_f}}\right)\left(\frac{\partial^2 T}{\partial x^2}+\frac{\partial^2 T}{\partial y^2}\right) \quad (2) \qquad \frac{\partial^2 \Psi}{\partial x^2}+\frac{\partial^2 \Psi}{\partial y^2}=-\Omega \quad (3)$$

The dimensionless tangential and normal velocities are converted to:

$$u=\frac{\partial \Psi}{\partial y}\ ;\ v=-\frac{\partial \Psi}{\partial x} \quad \text{and} \quad \Omega=\frac{\partial v}{\partial x}-\frac{\partial u}{\partial y} \quad (4)$$

Nanofluid effective density, heat capacity, thermal expansion coefficient, effective dynamic viscosity, effective thermal conductivity and thermal diffusivity are, respectively, calculated as follow:

$$\rho_{nf}=\phi\rho_s+(1-\phi)\rho_f \quad (5) \qquad (\rho c_p)_{nf}=\phi(\rho c_p)_s+(1-\phi)(\rho c_p)_f \quad (6) \qquad (\rho\beta)_{nf}=\phi\rho_s\beta_s+(1-\phi)\rho_f\beta_f \quad (7)$$

$$\mu_{nf}=\frac{\mu_f}{(1-\phi)^{2.5}} \quad (8) \qquad \frac{\lambda_{nf}}{\lambda_f}=\frac{\lambda_s+2\lambda_f-2\phi(\lambda_f-\lambda_s)}{\lambda_s+2\lambda_f+\phi(\lambda_f-\lambda_s)} \quad (9) \qquad \alpha_{nf}=\frac{\lambda_{nf}}{(\rho c_p)_{nf}} \quad (10)$$

Where "f", "s" and "nf" indicate fluid, solid particles and nanofluid respectively.

2.1. Boundary Conditions

The boundary conditions applied to the problem are written as:

$u=v=0$ on the rigid walls

$T=1+a\sin\left(\frac{2\pi x}{A}+\gamma\right)$ on the lower horizontal heated wall (a = 0.5 stands for the amplitude of the sinusoidal temperature)

$\frac{\partial T}{\partial n}=0$ on the adiabatic walls ("n" indicates the normal direction to the considered adiabatic wall)

$\Psi=0$ on the walls below the two openings

$\Psi=B$ on the walls above the two openings

$T=0$ at the inlet port

$u=1,\ v=0,\ \Psi=y-(1-B),\ \Omega=0$ at the outlet port

• Many extrapolation techniques are known for the last years in order to obtain the unknowns boundary conditions at the exit. Some technics are based on the incompressibility property of the fluid and others on the condition of an established flow. However, for a wide range of physical problems, the outlet flow has a complex flow structure. This led us to adopt an unrestrictive technique which consists of zero seconds derivatives at the outlet for the temperature. In the other hand, u, v, Ψ and Ω are unknown at the inlet. These quantities are also extrapolated at each time step by assuming nil their second derivatives at this port.

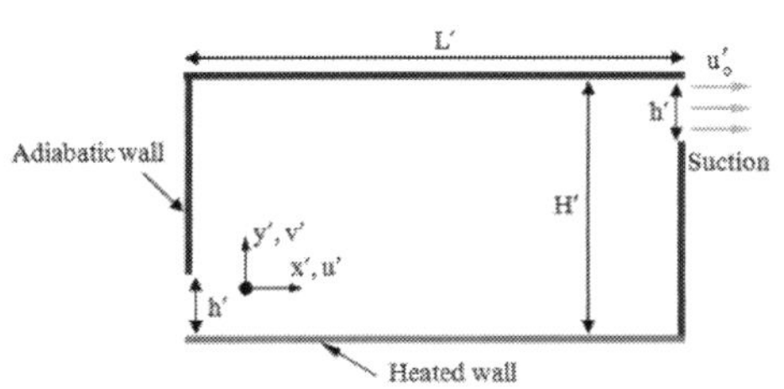

Figure 1. Geometry and coordinates system

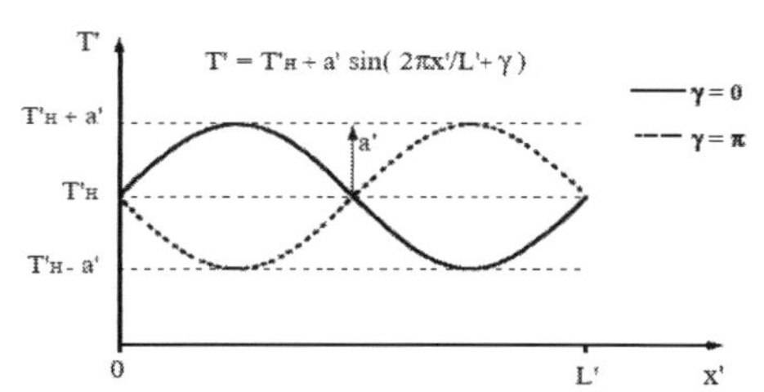

Figure 2. Sinusoidal temperature profile on the heated wall

2.2. *Heat transfer*

The estimation of the heat transfer enhancement is based on the average Nusselt number, Nu, calculated on the heated bottom wall of the cavity such as:

$$\mathrm{Nu} = -\frac{1}{A}\left(\frac{\lambda_{nf}}{\lambda_f}\right)\int_0^A \left.\frac{\partial T}{\partial y}\right|_{y=0} dx \tag{11}$$

3. Numerical approach

Governing equations for continuity, momentum and energy equations associated with the boundary conditions in this investigation were calculated numerically based on the finite difference method using FORTRAN computer code. The first and second derivatives of the diffusive terms were approximated by a second-order central difference scheme. Furthermore, the advection terms were handled using a second-order upwind differencing scheme to avoid possible instabilities frequently encountered in mixed convection problems. Then, an alternating direction implicit procedure (ADI) was used to perform the time integration for Eqs. (1) and (2). At each time step, the Poisson equation, Eq. (3), was treated by using the point successive over-relaxation method (PSOR) with an optimum over-relaxation coefficient equal to 1.95 for the grid (201 × 101) retained in this study.

4. Results and discussion

The Rayleigh number was kept at a constant value ($Ra = 10^6$) while the Reynolds number, Re, ranging between 200 and 5000. Values of these parameters involve values of the Richardson number, Ri, varying in the range [6.45×10^{-3}, 4.03] (with Pr = 6.2 for water). In the following, effects of Re, ϕ and γ on flow and thermal fields and heat transfer effectiveness are examined.

Comparisons between results of pure water ($\phi = 0$) and nanofluid ($\phi = 0.05$), expressed in terms of streamlines and isotherms, are displayed in Figs. 3a and 3b for moderate value of Re (Re = 700) and two values of γ. As shown in Fig. 3a, for heating with zero phase shift ($\gamma = 0$), the open lines of the forced flow are quasi-parallel to the heated wall in the entry area allowing more thermal interaction between the fluid and this wall and forcing the lower closed cell to be cloistered in a restricted area in contact with the right vertical wall. At the same time, the corresponding isotherms are tightened at the level of the heated horizontal wall, from the inlet to its center, indicating a good convective heat transfer, characterized by a limited increase of the thermal boundary layer thickness from the inlet toward the center of the lower wall. In another side, three-quarters of the cavity space is at the ambiance temperature. Note also that the addition of nanoparticles has a limited qualitative effect on the streamlines but their effect is more visible on the temperature distribution in the heated area. For $\gamma = \pi$ (Fig. 3b), the change of phase shift leads qualitatively to similar flow structure. The corresponding isotherms indicate a progressive growth/decay of the zone with high temperature gradients in the right/left side of the heated wall from right to left. Since the growth in the right side occurs at a faster rate than the decay in the left side, we assist to a heat rate uniformly distributed over the whole lower wall. This behavior results from the fact that a predominant forced convection in the left side is balanced by a predominant natural convection in the right part of the heated wall.

For Re = 5000, streamlines and isotherms, obtained for two values of γ, are presented in Figs. 4a and 4b, for both pure water and nanofluid. According to Fig. 4a, obtained for $\gamma = 0$, the dynamical structure shows that highest values of Re favor the closed natural convection cell, located in the right bottom corner, due to the rising assisting effect of natural and forced convections. Also, forced convection amplifies the closed cell located in the left top corner by reducing the main one surmounting the open lines. Remark that, even for large values of Re, the open lines of forced convection remain curved near the heated bottom wall. The isotherms indicate generally a slight increase of the thermal boundary layer, as a disadvantage of the cold zone, which testifies of the increasing effect of natural convection occasioned by the greater lower cell. On the other hand, while referring to Fig. 4b, for $\gamma = \pi$, keeps unchanged the flow structure but modifies the thermal one. It is noteworthy that, the dynamical field is almost insensitive to any change of ϕ or γ, for the high values of Re, as depicted in Fig.4.

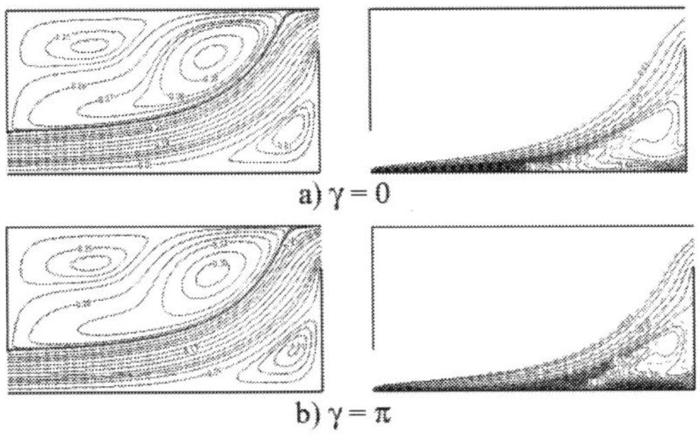

Figure 3. Streamlines and isotherms, for $\phi = 0$ (—) and $\phi = 0.05$ (- - -) for Re = 700 and different values of γ

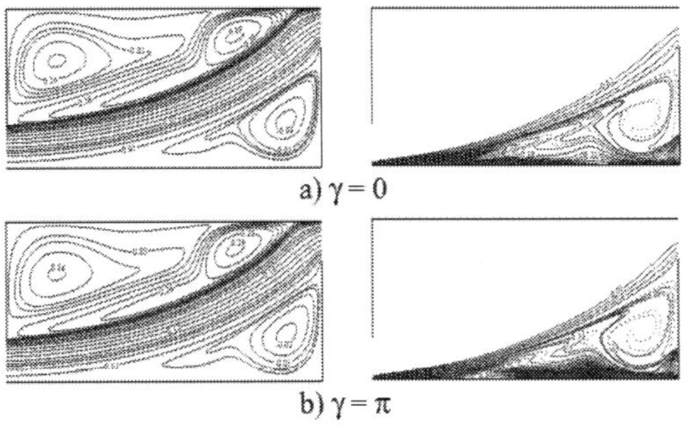

Figure 4. Streamlines and isotherms, for $\phi = 0$ (—) and $\phi = 0.05$ (- - -) for Re = 5000 and different values of γ

In order to illustrate the performance of the imposed flow mode on the process of heat removal, the variations of Nu, with Re, along the heated wall are presented in Fig. 5, for both heating temperature profiles (the case of periodic heating with zero phase shift ($\gamma = 0$) and $\gamma = \pi$ (temperature profile opposed to that corresponding to $\gamma = 0$)) and two values of ϕ. Overall, Nu increases monotonously with Re, for both cases of heating temperature profile, because of the flow intensification by the forced inertial effects. For a fixed value of Re, an augmentation of ϕ up to 0.05 affects positively the effective thermal conductivity of the nanofluid, which leads to a manifest increasing effect of convection, either for both heating modes. It should be noted that, the case of periodic heating without phase shift ($\gamma = 0$) enhances more the heat transfer in comparison with the other case of heating, either in case of a cavity filled with a base fluid (pure water) or with a nanofluid. These results allow concluding that a better heat removal from the heating wall is ensured by applying the first ($\gamma = 0$) case of heating mode.

For practical purposes, the evaluation of the nanofluid mean temperature, $\overline{T}$, inside the cavity is of a great interest. Hence, the variations of this quantity with Re are plotted in Fig. 6, for two values of ϕ and both heating temperature profiles. In the case of periodic heating with zero phase shift, the evolution of $\overline{T}$, with Re, is characterized by a weak decrease as long as Re is lower than a value around 500. Above this threshold value, the rate of decrease becomes more obvious. For the case of periodic heating with phase shift equal to π, the evolutions of $\overline{T}$ are characterized first by an increase with Re up to a critical value situated between 500 and 600 (depending on ϕ). This trend is attributed to the assisting flows of natural and forced convections due to the competition between them. Above this threshold the evolution is reversed and is characterized by a clear decrease of $\overline{T}$ with Re as a consequence of the forced flow predominance which favors the heat evacuation through the outlet. For a given value of Re, it can be observed that the addition of nanoparticles enhances notably $\overline{T}$ for both heating temperature profiles. Moreover, it is retained from this figure that the case of periodic heating with zero phase shift ($\gamma = 0$) involves globally to a better cooling inside the cavity since the related values of $\overline{T}$ are weaker in comparison with the other case of heating (for $\gamma = \pi$), if we except the case corresponding to weaker values of Re (Re $\leq$ 400) for which the case of periodic heating with phase shift ($\gamma = \pi$) gives the most cooling efficiency. The solutions obtained are unsteady periodic in the case of periodic heating with phase shift ($\gamma = \pi$), (full circles in Figs. 5 and 6) by increasing Re from its lower value up to a certain limit that depends on the volume fraction of nanoparticles.

Variation of Nu divided by λ_{nf}/λ_f with Re, presented in Fig. 7, is qualitatively similar to the evolution of Nu with Re (shown in Fig. 5). Quantitatively, a weak difference is observed when the naoparticles are added to the base liquid ($\phi \neq 0$). Precisely, for a fixed value of Re, $Nu/(\lambda_{nf}/\lambda_f) < Nu$.

Finally, it is to underline that a comparison between the case of a sinusoidal temperature heating and a uniform temperature heating was performed (Fig. 8). The findings showed that the heating by sinusoidal temperature profile is more favorable to the heat transfer.

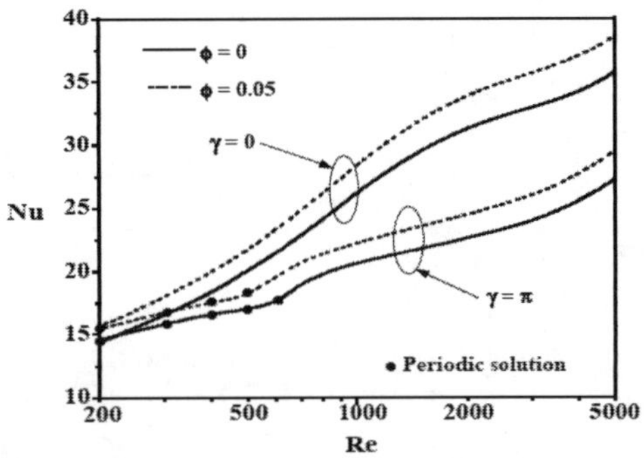

Figure 5. Variations, with Re, of the average Nusselt number, Nu, for both $\gamma = 0$ and $\gamma = \pi$ obtained for two values of ϕ.

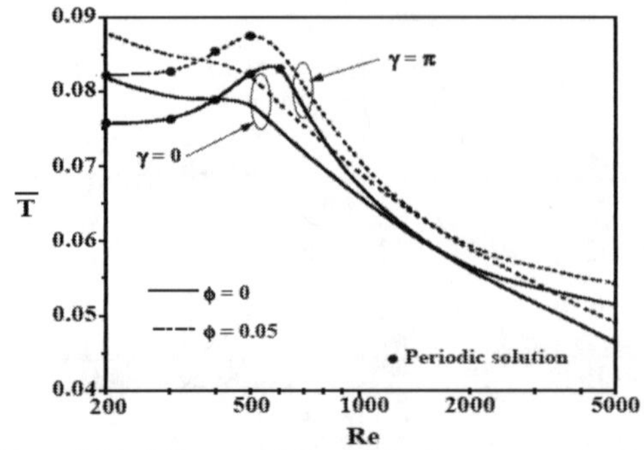

Figure 6. Variations, with Re, of the the mean temperature, $\overline{T}$, for both $\gamma = 0$ and $\gamma = \pi$ obtained for two values of ϕ.

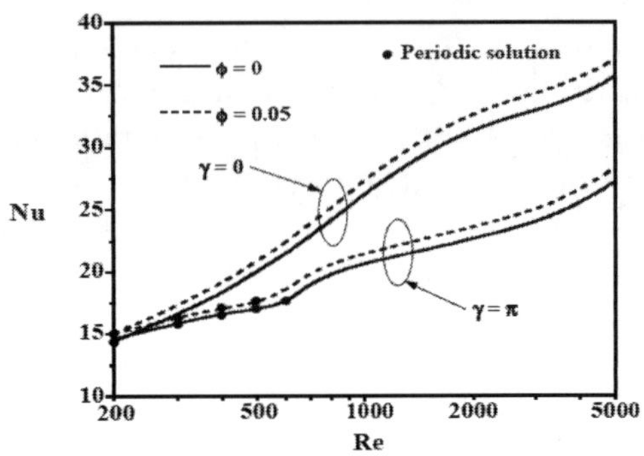

Figure 7. Variations, with Re, of the average Nusselt number, Nu, divided by (λ_{nf}/λ_f), for various values of γ and ϕ.

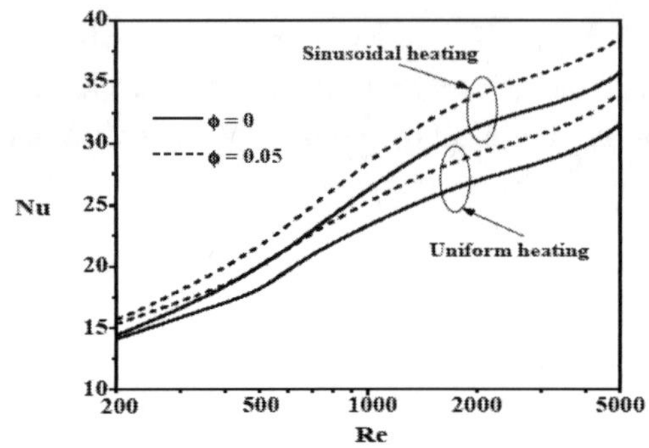

Figure 8. Variations, with Re, of Nu, for both sinusoidal and uniform heating obtained for two values of ϕ.

5. Conclusion

In this paper, a numerical study is carried out in order to investigate the effect of Al_2O_3-water nanofluid on mixed convection heat transfer in a vented rectangular cavity provided with two openings and submitted to a sinusoidal heating from below. The study focused on the suction mode of imposed external flows and two types of heating (periodic heating with or without phase shift). The obtained results indicate that Re affect significantly the thermal characteristics and the flow filed, which are found more sensitive to any change in ϕ and γ only for small values of Re. It is also seen that the presence of the nanoparticles in the base fluid acts to increase heat transfer rate and mean temperature inside the enclosure. On the other hand, it is to remember that the case of periodic heating without phase shift gives a better thermal efficiency than the case with phase shift ($\gamma = \pi$) by leading to more heat transfer across the cavity even with or without presence of nanoparticles. In addition, it is found that a better cooling of the cavity is generally reached by applying the case of periodic heating without phase shift (for moderate and high values of Re) and the case of periodic heating with phase shift (for low values of Re) since these cases lead to lower values of the mean temperature.

References

[1] Jana S, Khojin A S and Zhong W H 2007 *Thermochim. Acta* **462** 45-55.

[2] Arani A A A, Sebdani S M, Mahmoodi M, Ardeshiri A and Aliakbari M 2012 *Superla.t and Microstr.* **51** 893-911.

[3] Kefayati G H R 2014 *Powder Technology* **266** 268-281.

[4] Goodarzi M, Safaei M R, Vafai K, Ahmadi G, Dahari M, Kazi S N and Jomhari N 2014 *Inter Jour of Ther Scien* **75** 204-220.

[5] Karimipour A, Esfe M H, Safaei M R, Semiromi D T, Jafari S and Kazi S N 2014 *Phys A: Sta Mec and its App* **402** 150-168.

[6] Shahi M, Mahmoudi A H, and Talebi F 2010 *Int Com in Heat and Mass Transfer* **37** 201-213.

[7] Mahmoudi A H, Shahi M, and Talebi F 2010 *Int Com in Heat and Mass Transfer* **37** 1158-1173.

[8] Sourtiji E, Hosseinizadeh S F, Gorji-Bandpy M and Ganji D D 2011 *Int Com in Heat and Mass Transf* **38** 1125-1134.

IOP Conf. Series: Journal of Physics: Conf. Series **1226** (2019) 012002 doi:10.1088/1742-6596/1226/1/012002

Developing instrumentation for analysis of CPV TwinFocus® system

M Nardello[1,2,a] **and S Centro**[1,2,3,b]

[1]Museo Storico della Fisica e Centro Studi e Ricerche "E. Fermi", piazza del Viminale 1, Roma, Italy.
[2]Università degli studi di Padova, via 8 febbraio 1848, Padova, Italy.
[3]AtemEnergia srl, via Padania 19, San Vendemiano, Italy.

[a)]nardello@atemenergia.it
[b)]centro@atemenergia.it

Abstract. Concentrator photovoltaics (CPV) is a technology that offers an alternative to standard silicon modules, being more efficient and eco-friendlier. AtemEnergia is developing a CPV product with good characteristics of efficiency and reliability. To monitor the progresses of our prototypes, test their quality and discover potential defects in the devices and in the production lane, custom experimental setups were necessary, to investigate the peculiar characteristics of this technology, avoiding the huge costs of commercial instrumentation in this field. Here is an overview of the tests necessary to guarantee the correct production quality, along with the custom solutions we adopted for their implementation.

1. Introduction

Photovoltaic is the most interesting among renewable energy production technology, because of the abundance of the energy source and other strong points like long lifetime, low maintenance, ease of installation, no fuel requirement [1]. It has, however, some drawbacks, like the not optimal efficiency of the cells, which increase the area of installations and the costs of the product, the lack of energy production at night and in cloudy weather, the costs associated with used silicon treatment that rend impossible the recycling of spent panels.

CPV overcome some of these problems, because it avoids the use of silicon, preferring the more efficient multijunction solar cells. Standard silicon solar cells have a theoretical efficiency limit of 33.16% [2] that is due to the fact that light absorption is impossible for energies lower than the band gap of the material and conversion to electricity is rendered inefficient due to competitive relaxation mechanisms at energy higher than that of the band gap. But using different junctions made of different materials stacked together we can modulate the band gap so that each junction absorbs with minimal losses a different part of the spectrum, thus reaching a theoretical efficiency limit of 86.8% [3]. For practical application CPV cells have reached 46% record efficiency, rapidly increasing towards the 50% goal, against 27% of Si. Commercially available cells report 44% efficiency against standard silicon cells around 20% [4].

The superiority of these types of cell is given by another important characteristic: their efficiency is less affected by operation at high temperature. While for standard Si cells we note a decrease of 0.4%/°C efficiency for every degree of deviation from standard conditions (25°C), for multijunction solar cells the efficiency drop is only 0.04%/°C [5]. This allows for concentration of light in smaller areas,

IOP Conf. Series: Journal of Physics: Conf. Series **1226** (2019) 012002 doi:10.1088/1742-6596/1226/1/012002

decreasing drastically the amount of semiconductor used. Typical concentration factors for CPV go in fact from a few suns (LCPV) to more than 1000x (HCPV). The main structures of CPV modules are not composed by the "active" material, but by optical elements, lenses or mirrors, capable of reflecting light and concentrate it on the small cells. Optical materials used are glass, aluminum, plastics, very cheap and recyclable.

The correct alignment of the optics is guaranteed by moving structures, which follow the sun throughout the day. This characteristic maximizes the amount of produced energy, since solar rays strike the surface of the panels perpendicularly also in the morning and late afternoon, when efficiency of standard fixed panels decrease instead. The gain in produced energy provided by tracking can reach +20%, as shown by measurement made by us, independently by the type of cells used (figure 1).

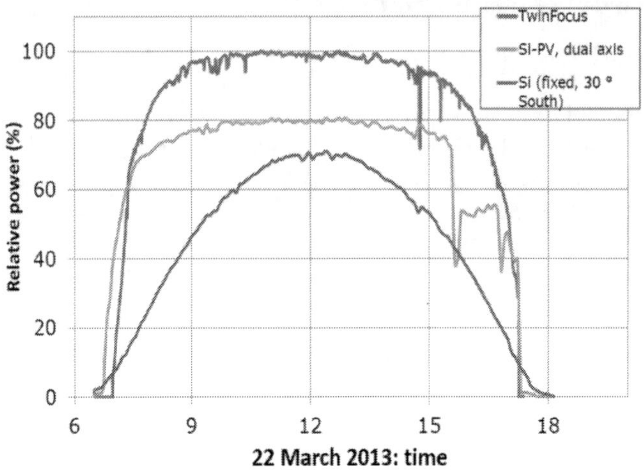

Figure 1. Graph showing measurement of produced power throughout the day for a fixed silicon panel, a silicon panel mounted on a tracking system, a prototype of our CPV system TwinFocus (first version). Measurement are normalized to the peak power production of TwinFocus system.

The properties just described make CPV a favorable technology for energy production, especially in dry and sunny areas, where scattered light is not predominant and there is high DNI (direct normal radiation). But the intrinsic complexity of these type of structure makes it less competitive on the cost issue; for this reason, CPV systems developers have to pose a great deal of attention in research both to develop a design that limit the costs and to maintain the high-quality standard that is the counterpart of the greater initial investment.

2. TwinFocus®

It is in this context that in 2010 began the collaboration between University of Padova, Centro Fermi, and a group of private companies, operative in the design and production of molds and plastic products, for the realization of a CPV product, with a custom optical design and the ambition to become a reliable and competitive solution for energy production.

The resulting product is named TwinFocus®. It saw a series of modifications and improvement from its original design to arrive nowadays at a 31% efficiency at concentrator standard test conditions (CSTC). Its design has already been described elsewhere [6], so here we will show just a brief illustration.

2.1. Opto-electrical unit

The core of the system is based on a curved mirror, divided in two halves that deviate and focus the solar light on two receiver assembly, located on the sides of the optical elements (see figure 2 and 3).

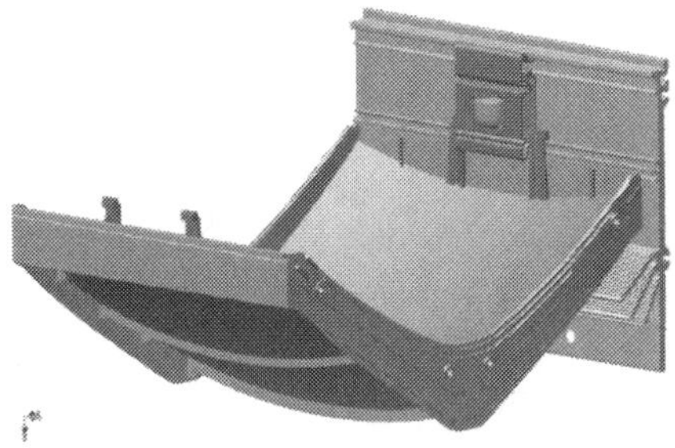

Figure 2. The mirror is shown, it is connected on the right with the aluminium bar that act as a support. On the same bar is glued the receiver

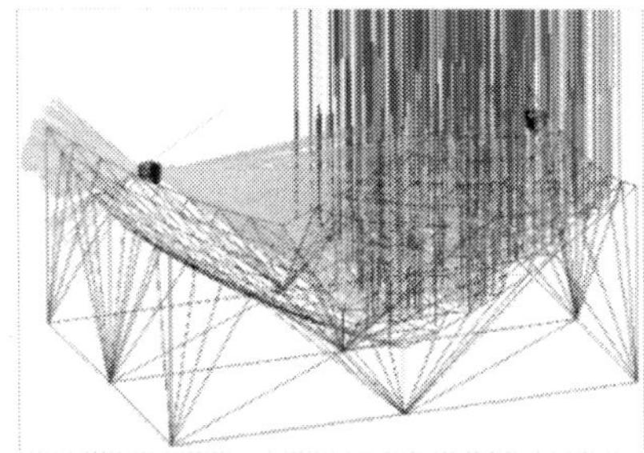

Figure 3. Schematic representation of the optical element, the light path is shown for the right half of the mirror.

2.1.1. Primary optics. The mirror was designed and realised internally. The design is off-axis, with each half of the mirror composed by 6 quadrants of parabolic shape, focusing on different points of the receiver. It is realised in polycarbonate coated with a reflective layer of aluminium.

2.1.2. Receiver assembly. The light deflected by the mirror strikes first on a secondary optic, a silica prism that funnels it, thanks to total internal reflections, towards the solar cell located at its base. The solar cell is an Azur Space 3C44C [7], a triple junction solar cell with nominal efficiency of 44% and dimensions 5.5x5.5 mm2 to attach the cell to the wall of our modules is necessary a substrate that eliminates thermal stresses and provides the electrical connection. This is a DBC (direct bonded copper) made of a layer of alumina sandwiched between two copper foils.

2.2. Structure and installation

The units described in *§2.1* are linked together in a series of 12 and enclosed in an aluminum structure, that provides stability, water tightness and thermal dissipation. The overall design is lightweight, simple, and allows for easy electrical connections among cells and to the outside. An UV transparent glass covers the module on top. Figure 5 shows the design.

The modules are mounted on a tracking system. Our choice is that of a pole installation, that allows for passaging underneath and keeping the land available for other uses, such farming, parking etc minimizing the impact of the structure on the surrounding environment and its activities. Figure 4 show our first prototype installation, on which we are now collecting the first efficiency and power production data.

Figure 4. TwinFocus tracker 3.6kW installation

IOP Conf. Series: Journal of Physics: Conf. Series **1226** (2019) 012002 doi:10.1088/1742-6596/1226/1/012002

3. Testing activities

A fundamental part in the development of a new product is the analysis of its performances and the overall quality of the process. CPV has characteristics that differentiate it from classic photovoltaic and thus need different instrumentation especially designed for its testing. This instrumentation can be not easily available for custom devices such ours, and often very expensive. We realised optical and electrical setups specifically tailored for our needs and used them to monitor every step of the creation of our modules.

3.1. Electrical and optical performances tests

Figure 5. Experimental setup, Xe lamp in the centre with power supply (up to 300W), mechanical rotating support for the cell on the optical bench, active load on the right, NI hardware on the optical bench. The substrate is rotated at the angle that maximise the absorption of light from the cell.

Solar cells are current generators, whose power production is dependent on the electrical load applied to them. They present a characteristic curve of produced current as a function of the applied voltage which is comparable with the characteristic of a diode [8]. If no load is applied to the cell, the produced current is called short circuit current (I_{SC}), increasing the voltage to the cell the current decreases until become zero at the open circuit voltage (V_{OC}). The maximum power that can be obtained from the cell is P_M; defects in the cell can lower P_M modifying the shape of the curve.

Looking at the IV curve is a good method to test the quality of a solar cell. To do so we realised a setup composed by a Xe lamp (with a spectrum similar to that of the Sun), a plano-convex lens to focus the emitted light on the cell, and a support for the cell providing electrical connections and a variable load able to generate a voltage ramp applied to the cell and to collect the produced current as a function of the voltage (figure 5). The obtained signal is analysed by National Instrument hardware and the characteristic values are calculated thanks to a program created in Labview (figure 6).

The Xe lamp is useful not only for electrical tests of the cell but also to check the performances of the optical elements. With a system of optical fibres, filters, diffusers, beam splitters, a photodiode and a spectrophotometer, we can check the punctual reflectivity of mirrors and the transmittance of glass and secondary optics.

3.2. Thermal performances tests

An aspect that is peculiar to CPV is the great importance of thermal management, since concentrating light brings a lot of power in small areas; power that has to be dissipated efficiently if we want the cell to work at optimal conditions. For this reason, the choice of the glue was made after careful examination of the available options. Sadly, values of thermal conductivity of various products are obtained with different techniques and are not trustworthy.

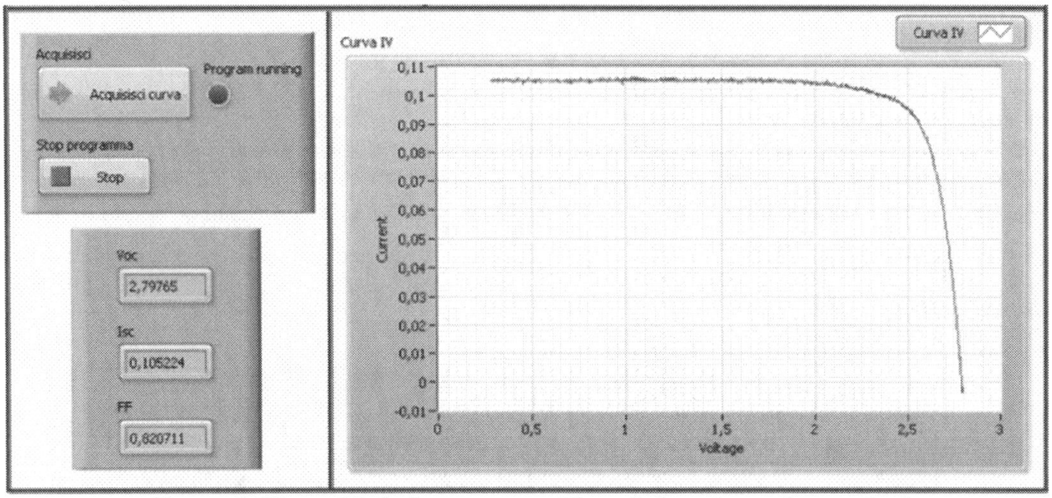

Figure 6. View of Labview program. The window shows the acquired I-V curve (blue line on the right), highlighting (bottom left) key parameters as V_{OC}, I_{SC} and FF. Units of measurement are volts for voltage and ampere for currents. FF is the fill factor, a quality check parameter obtained as the ratio between the maximum obtained power and the maximum theoretical power given by $V_{OC}*I_{SC}$

To overcome this problem, we devised a temperature testing setup to directly measure the temperature of the cell. The increase of temperature is provided applying a current through the cell (the cell acts as a LED) of 4A. The cell is glued with the product under test on its substrate and to an aluminium heat sink. The heat sink is kept at a fixed temperature by an active cooling system (a metal block with channels of flowing water cooled at 10°C). The difference of temperature between cell and substrate reach at equilibrium a constant value dependent on the conductivity of the glue. Applying a resistance thermometer to the cell we can easily calculate the best thermal glue (figure 7).

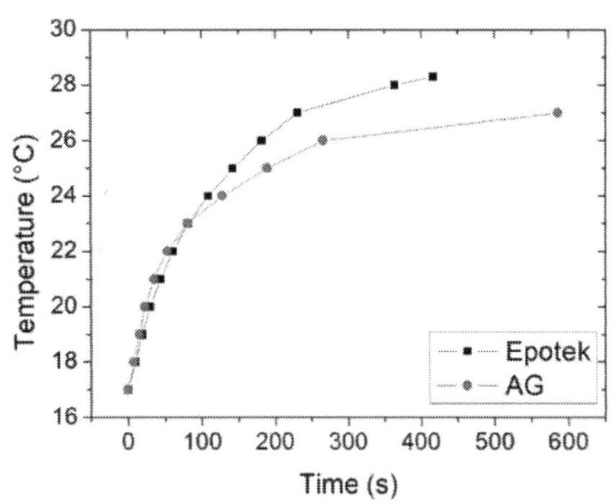

Figure 7. Comparison between two commercial glues. The cell shows a smaller increase in temperature when AG glue is applied and reach a lower equilibrium temperature when current (4A) is applied for a long period of time.

3.3. *Visual quality checks*

Applying current to the cell is a good way to spot other defects in the device. Since the cell under these conditions emits light (photoluminescence), the observation of the emitted light shows details not evident observing the cell with the naked eye.

Of particular importance is the intensity of the emitted radiation because points in the cell may present slightly different electrical conduction properties, favouring the passage of current and increasing locally the temperature (hot spots). This in turn increase the conductivity and establish a cascade effect that can bring to the rupture of the cell (figure 8).

Electroluminescence emission shows these defects as brighter spot, allowing to recognise and isolate the faulty cells.

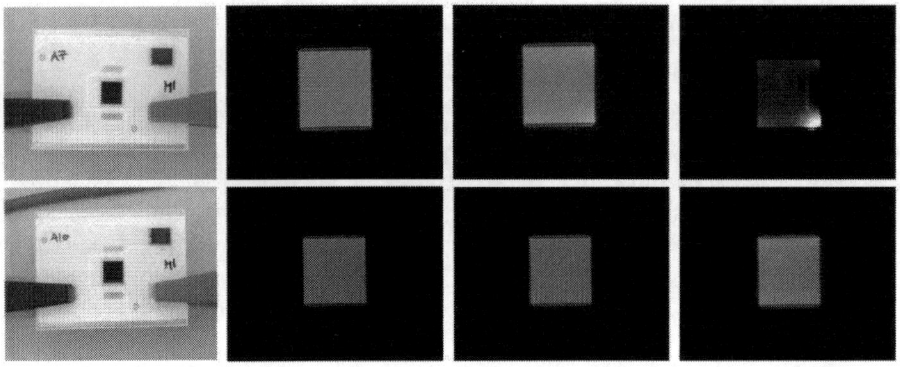

Figure 8. Electroluminescence test for two cells. In the top row a brighter spot is visible on the bottom right corner. Increasing the current (from left to right) the difference become more visible, until the cell breaks (picture on the right). The bottom row shows a cell without defects.

4. Conclusions

The project TwinFocus is a great example of collaboration between industry and university. It allowed for developing a fully mature and competitive technology that can produce energy with great efficiency. It involved a good amount of research, of which I have shown here some example relative to the development of custom instrumentation.

Of particular importance were the realization of electronic devices for performance testing and the setup for the thermal management of the cell.

References

[1] Shayestegan M et al. 2018 *Renewable and Sustainable Energy Reviews* **82(1)** 515–30
[2] Rühle S 2016 *Solar Energy* **130** 139–47
[3] Green M A 2003 *Third Generation Photovoltaics: Advanced Solar Energy Conversion* Springer 65
[4] Philipps S 2018 *Photovoltaics report* Fraunhofer ISE and Werner Warmuth, PSE Conferences & Consulting GmbH
[5] Simone M. 2013 Fotovoltaico a Concentrazione: le Celle Multigiunzione *Ph.D. thesis, University of Padova*
[6] Nardello M and Centro S 2017 *AIP Proc. Conf. CPV13* **1881** 030006 10.1063/1.5001417
[7] http://www.azurspace.com/images/products/0004356-00-01_3C44_AzurDesign_5.pdf
[8] Hovinen A 1994 *Physica Scripta* **1994(T54)** 175 –6

Rotation in galaxy clusters from MUSIC simulations with the kinetic Sunyaev–Zel'dovich effect

A S Baldi[1], M De Petris[1], F Sembolini[1,2], G Yepes[2], W Cui[2], L Lamagna[1]

[1] Dipartimento di Fisica, Sapienza Università di Roma, Piazzale Aldo Moro 5, I-00185 Roma, Italy

[2] Departamento de Física Teórica, Universidad Autónoma de Madrid, Calle Francisco Tomás y Valiente 7, E-28049 Madrid, Spain

E-mail: annasilvia.baldi@uniroma1.it

Abstract. The Sunyaev–Zel'dovich effect in galaxy clusters is a unique probe for studying astrophysics and cosmology. We propose in this work its application for the detection of possible coherent rotational motions in the hot intra-cluster medium. We select a sample of massive, relaxed and rotating galaxy clusters from Marenostrum-mUltidark SImulations of galaxy Clusters (MUSIC), and we produce mock maps of the temperature distortion produced by the kinetic Sunyaev–Zel'dovich effect by exploring six different lines of sight, in the best observational condition. These maps are compared with the expected signal computed from a suitable theoretical model in two cases: (*i*) focusing only on the contribution from the rotation, and (*ii*) accounting also for the cluster bulk motion. We find that the parameters of the model assumed for the radial profile of the rotational velocity, averaged over the considered lines of sight, are in agreement within two standard deviations at most with independent estimates from the simulation data, without being significantly affected by the presence of the cluster bulk term. The amplitude of the rotational signal is, on average, of the order of 23 per cent of the total signal accounting also for the cluster bulk motion, and its values are consistent with the literature. The projected bulk velocity of the cluster is also recovered at the different lines of sight, with values in agreement with the simulation data.

1. Introduction

The study of non-random motions in the diffuse gas in galaxy clusters (the intra-cluster medium, or ICM) can play an important role for precision cosmology. Indeed, cluster masses are a valuable observable for constraining some cosmological parameters, and accurate mass measurements should take into account the contribution from possible ordered motions, which would be given by an effective non-thermal pressure gradient [1]. Recent results from hydrodynamical simulations have shown, indeed, that mass estimates accounting only for the support from thermal pressure are generally biased towards values 10-20 per cent lower than the true ones (see e.g. [2] and references therein). With the aim of quantifying the non-thermal correction to the hydrostatic mass, several approaches have been proposed to detect, in particular, possible ordered *rotational* motions in the ICM, through observations of the ICM itself or of the galaxy members at X-ray [3], millimetre [4–6], and optical wavelengths [7–9].

IOP Conf. Series: Journal of Physics: Conf. Series **1226** (2019) 012003 doi:10.1088/1742-6596/1226/1/012003

In this work, we focus on the detection of a possible rotation of the ICM based on measurements in the microwave band of the kinetic component of the Sunyaev–Zel'dovich (SZ) effect [10, 11]. The SZ effect arises from the Comptonization of the photons of the cosmic microwave background (CMB), as they propagate through galaxy clusters and interact with the free electrons in the ICM. The scattering causes the redistribution of the energy of the photons, which can be measured as an intensity (or temperature) distortion of the CMB blackbody spectrum. Two components can be distinguished in the SZ effect: one is due to the thermal motion of the electrons (the thermal SZ effect, tSZ); the other one is produced by the motion of the cluster as a whole with respect to the CMB rest frame (the kinetic SZ effect, kSZ). Since it is proportional to the ICM velocity, the kSZ can be a useful probe of cluster dynamics. In particular, a purely rotational motion of the ICM would be detectable from the kSZ temperature distortion, which is expected to feature a dipole pattern, due to the projected approaching and receding velocities of the rotating gas relatively to the observer, as investigated in [4] and [5] for the first time.

Synthetic galaxy clusters from large cosmological hydrodynamical simulations represent ideal targets for preliminary studies on the feasibility of the detection of a rotation with the aforementioned possible observational strategies. The MUSIC[1] catalogue [12] is one of the largest data set of simulated galaxy clusters to date. We select six candidate rotating clusters from MUSIC according to the results we reported in a companion paper [13], and we produce numerical maps of the kSZ temperature signal. In order to constrain their rotation, we fit them to the expected theoretical kSZ maps for a rotating ICM. In particular, we assume the model by [4] as a starting basis, and we update it by accounting for a different law describing the rotational velocity we proposed in [13], assuming both a rotation-only case, and the full case accounting also for the cluster bulk motion.

The paper is organized as follows. Sec. 2 briefly reports the features of our cluster data set. In Sec. 3 we describe the theoretical and mock kSZ temperature maps. In Sec. 4 we report our results, and eventually we summarize our conclusions in Sec. 5.

2. Data set

We select our target clusters from the MUSIC-2 catalogue which is based on high-resolution re-simulations of clusters from the MultiDark parent simulation [14], from which we also inherit the assumed cosmological model (see [12] for details). Simulations are run with the GADGET-3 Tree-PM+SPH (i.e. using particle-mesh and smoothed particle hydrodynamics) code, and follow the evolution of gas and dark matter particles, which have masses $m_{\rm gas} = 1.9 \times 10^8 h^{-1}{\rm M}_\odot$ and $m_{\rm DM} = 9.0 \times 10^8 h^{-1}{\rm M}_\odot$, respectively (where ${\rm M}_\odot$ is the mass of the sun and $h = 0.7$). For our purposes of studying rotation, we focus in particular on a sample of massive (virial masses $M_{\rm vir} > 5 \times 10^{14} h^{-1}{\rm M}_\odot$), dynamically relaxed objects at redshift $z = 0$, with spin parameter of the gas $\lambda_{\rm gas} > 0.07$ (see the companion work by [13] for further details). The main properties of our cluster sample selected according to these criteria are summarized in Tab. 1. We limit our present analysis to objects extracted from the adiabatic run, i.e. without accounting for radiative baryon physics processes, since they do not have significant impact on the rotational properties of the gas [13].

[1] `http://music.ft.uam.es`

IOP Conf. Series: Journal of Physics: Conf. Series **1226** (2019) 012003 doi:10.1088/1742-6596/1226/1/012003

Table 1. Identifier, virial mass, virial radius and spin parameter of the gas of the six galaxy clusters analysed in this work.

cluster ID	$M_{\rm vir}$ ($\times 10^{15}$ M$_{\odot}$)	$R_{\rm vir}$ (kpc)	$\lambda_{\rm gas}$
46	1.17	2756	0.0785
93	1.90	3241	0.0769
98	1.61	3071	0.0735
103	1.02	2633	0.0746
205	1.24	2813	0.0763
256	1.31	2867	0.0714

3. Maps of the kSZ effect

3.1. Theoretical model maps

The map of the kSZ temperature distortion produced by the rotation of the ICM described by a generic angular velocity law $\omega(r)$, can be written as [4]:

$$\frac{\Delta T_{\rm kSZ,r}(R,\phi)}{T_{\rm CMB}} = -\frac{\sigma_T}{c} R\cos\phi\sin i \int_R^{R_{\rm vir}} n_e(r)\;\omega(r)\;\frac{2r\;dr}{\sqrt{r^2-R^2}}\,, \tag{1}$$

which corresponds to a dipole. Here (R,ϕ) are the polar coordinates on the map, $T_{\rm CMB} = (2.725 \pm 0.001)$ K is the CMB monopole temperature [15], σ_T is Thomson cross section, c is the speed of light, and i is the angle between the line of sight and rotation axis of the ICM. The best observational configuration i.e. the one at which the rotational signal is maximum (corresponding to $i = 90°$), is represented schematically in Fig. 1. It can be seen from Eq. (1) that the two relevant physical quantities in the signal are the radial profiles of the angular velocity, $\omega(r)$, and of the electron number density, $n_e(r)$. To disentangle their contribution to the signal, the electron number density has to be estimated from independent data (e.g. from X-ray photometry). In our case we use the numerical profiles of $n_e(r)$ computed from the simulation data for each cluster, and we fold the corresponding best-fit analytical profiles into Eq. (1). To derive these profiles we fit our data to a simplified six-parameter Vikhlinin model [16]

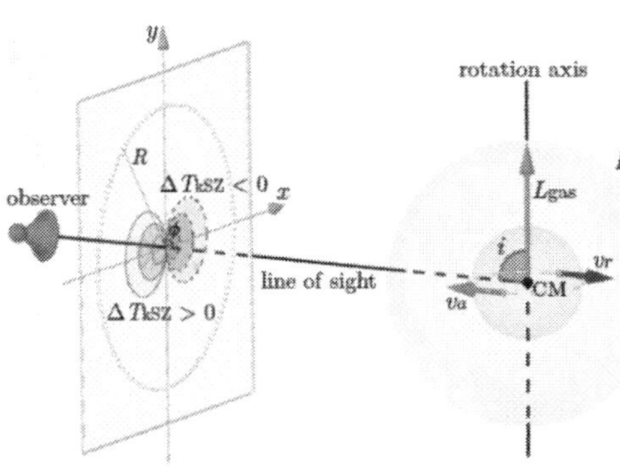

Figure 1. Schematic view illustrating the expected kSZ map from a rotating cluster, assuming the best observational configuration (i.e. with the line of sight orthogonal to the axis of rotation). The gas distribution in the cluster is assumed to be spherically symmetric, with the rotation axis aligned with the angular momentum vector of the gas ($\boldsymbol{L}_{\rm gas}$). The velocity vectors, $\boldsymbol{v}_a$ and $\boldsymbol{v}_r$, indicate respectively the approaching and receding velocity components along the line of sight for two generic gas particles, located at the same radial distance from the centre of mass of the cluster.

IOP Conf. Series: Journal of Physics: Conf. Series **1226** (2019) 012003 doi:10.1088/1742-6596/1226/1/012003

through a Markov chain Monte Carlo procedure. The angular velocity profile instead, is derived from the so-called vp2b model for the tangential velocity proposed in [13], as follows:

$$\omega(r) = \frac{\mathrm{vp2b}(r)}{r} = \frac{v_{t0}/r_0}{1 + (r/r_0)^2} , \tag{2}$$

therefore $\omega(r)$ depends on the two free parameters, r_0 and v_{t0}, which we want to recover from the fit to the mock kSZ maps.

When the cluster bulk motion is taken into account, and also a rotation is present, the total kSZ signal becomes:

$$\frac{\Delta T_{\mathrm{kSZ}}(R,\phi)}{T_{\mathrm{CMB}}} = \frac{\Delta T_{\mathrm{kSZ,r}}(R,\phi)}{T_{\mathrm{CMB}}} - \frac{\sigma_T}{c}\, v_{\mathrm{bulk}} \int_R^{R_{\mathrm{vir}}} n_e(r)\, \frac{2r\, dr}{\sqrt{r^2 - R^2}} , \tag{3}$$

where $\Delta T_{\mathrm{kSZ,r}}(R,\phi)/T_{\mathrm{CMB}}$ is given in Eq. (1), and v_{bulk} is the projection of the cluster bulk velocity on the line of sight This additional term produces an asymmetry in the dipole signal given by the rotation, depending on the dominating approaching or receding contribution from v_{bulk}.

3.2. Synthetic maps

We use the **pymsz**[2] code to produce the mock maps of our clusters. To compute the maps we simply discretize the signal given by Eq. (3), by summing over the velocity of the gas particles along a given line of sight times the electron density, properly accounting for the projection of the smoothing kernel used in the simulation. To distinguish the rotation-only case, we subtract the average velocity of the gas particles from the single particle velocity. The maps extend over $2.5R_{\mathrm{vir}}$ on each side, with pixel sizes of the order of 10 kpc. The angular size subtended by each pixel is $\sim$ 10 arcsec, which we degrade to a common resolution of 20 arcsec by means of a Gaussian smoothing, in order to emulate observations with a currently operating instrument (e.g. NIKA2 at $\sim$ 200 GHz).

To validate the rotational origin of the dipole signal in our mock kSZ maps, we explore six lines of sight orthogonal to the rotation axis (i.e. fulfilling the condition $i = 0$), at constant steps of 30°; we label each one of them with the value of the angular separation, θ_{los}, with respect to the reference line of sight (at $\theta_{\mathrm{los}} = 0°$). The dipole produced by rotation is expected to remain unchanged across these different projections, as well as the parameters describing it.

4. Results and discussion

With the aim of recovering the free parameters of the vp2b profile of the tangential velocity introduced in Eq. (2), we calculate a pixel-to-pixel fit to the mock kSZ maps using the two models given by equations (1) (for the rotation-only case) and (3) (for the general case). Fig. 2 illustrates the kSZ maps in the case of cluster 93 (the most massive in the sample), without (Fig. 2a) and with (Fig. 2b) such bulk contribution. Odd columns show the data maps, while even columns show the corresponding best-fit maps. It can be seen that the overall features of the signal are well recovered by our procedure at all the projections. It is also evident how the dipole becomes asymmetric across different lines of sight in the complete case, because of the dominating contribution from the projection of the approaching bulk velocity with respect to the observer. The small-scale outliers in the maps – which are characterized by a signal up to a factor of $\sim$ 3 higher than the best-fit signal – are produced by small high-velocity substructures encountered along the line of sight. Their presence, however, does not affect significantly the

[2] `https://github.com/weiguangcui/pymsz`

IOP Conf. Series: Journal of Physics: Conf. Series **1226** (2019) 012003 doi:10.1088/1742-6596/1226/1/012003

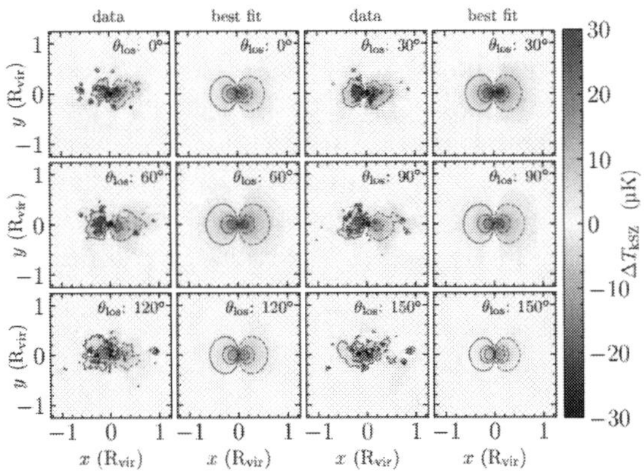

(a) without bulk motion

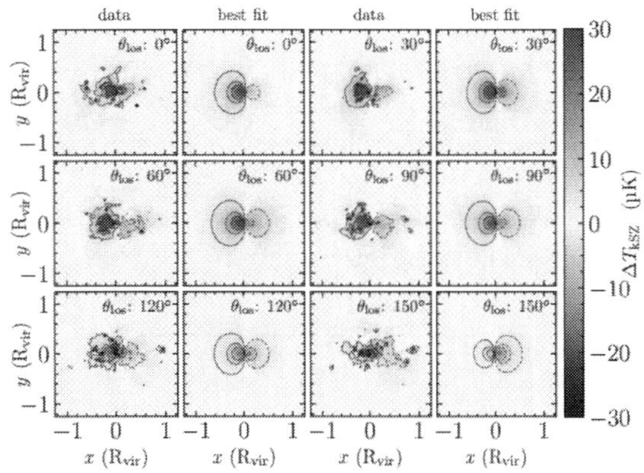

(b) with bulk motion

Figure 2. Mock maps of the kSZ temperature distortion for cluster 93, and corresponding best-fit maps, without and with accounting for the cluster bulk velocity (left and right panels, respectively). Contours are plotted from -5σ to 5σ, with dashed (solid) lines for negative (positive) values.

detection of the dipole in terms of amplitude or spatial features. The values of the two free parameters of the rotational model we recover from the fit to the kSZ maps of our cluster sample – r_0 and v_{t0} – are averaged over the explored line of sight. From the comparison of the rotation-only and the rotation+bulk cases, we find agreement within one standard deviation for both parameters. When we compare them to the values from an independent fit to the velocity data from the simulation instead, we get an agreement within at most two-standard deviations for v_{t0}, which is generally overestimated. This could be due to a higher sensitivity of this parameter to the presence of high-velocity outliers located in the outermost regions, or to a non-optimal reconstruction of the signal due to irregularities in the gas density. These results indicate that the vp2b model is, in general, a fairly good description of the tangential velocity profile, and that it is possible to separate the rotational contribution to the signal from the one coming from the bulk motion. This aspect is further confirmed by the lack of significant mutual correlation between the three free parameters, $v_{\rm bulk}$, r_0 and v_{t0}.

We quantify the rotational signal through the amplitude of the dipole, $A_{\rm dip}$, measured in the best-fit maps for the rotation-only case. The average values across the different lines of sight are listed in the middle column of Tab. 2: they are of the order of few tens of μK, in agreement with estimates from the literature for relaxed clusters [5]. The right column of Tab. 2 shows the amplitude $A_{\rm bulk}$ of the total kSZ signal given by rotation and bulk, as measured in the

Table 2. Amplitude of the kSZ temperature signal measured from the best-fit maps in the rotation-only case and in the rotation+bulk case (see text).

cluster ID	$A_{\rm dip}$ (μK)	$A_{\rm bulk}^{\rm max}$ (μK)
46	10.8 ± 2.5	-57.5
93	21.1 ± 5.2	82.1
98	24.4 ± 9.2	-77.9
103	16.5 ± 2.9	-99.4
205	20.9 ± 5.1	68.8
256	24.4 ± 4.3	-143.1

corresponding best-fit map at the line of sight in which $v_{\rm bulk}$ takes its maximum absolute value. From a comparison between these amplitudes it can be seen that, on average, $A_{\rm dip}/A_{\rm bulk}^{\rm max} \sim 0.23$, therefore a possible measurement of the rotational signal requires high sensitivity, coupled to a sufficiently high angular resolution to detect the dipole at sub-arcminute scales.

To validate our reconstruction of the projected bulk velocity of the clusters, $v_{\rm bulk}$, we compare the values derived from the fit at different projections with the true values from the simulation data. We find relative differences of the order of few tens per cent between the two at most projections; this indicates that in principle our procedure can be used to constrain the component along the line of sight of the cluster velocity field.

5. Conclusions

We tested the application of the kSZ effect for the detection and the characterization of possible rotational motions in the ICM, using a selected sample of massive and relaxed objects extracted from MUSIC simulations. Assuming the best observational configuration, we computed the synthetic and the theoretical maps of the kSZ temperature distortion assuming only a rotational motion, and also adding the contribution from the cluster bulk motion. In both cases we find a good agreement with the expected model for the rotational velocity, which we also test using the velocity data extracted from the simulations. The rotational signal is found to contribute about 23 per cent of the total maximum kSZ signal, on average, from the study of different lines of sight. Also, the projected cluster bulk velocity and the amplitude of the rotational signal are found to be consistent with the simulation values and with literature expectations, respectively. We are planning to extend this analysis by probing the rotation of ICM in our cluster sample also by means of complementary probes at different wavelengths, and by accounting for instrumental effects by referring to a specific candidate experiment for microwave astronomy.

Acknowledgements

ASB acknowledges funding from Sapienza Università di Roma - Progetti per Avvio alla Ricerca Anno 2017, prot. AR11715C82402BC7.

References

[1] Biffi V, Dolag K and Böhringer H 2011 *Mon. Not. R. Astron. Soc.* **413** 573–584
[2] Biffi V *et al.* 2016 *Astrophys. J.* **827** 112
[3] Bianconi M, Ettori S and Nipoti C 2013 *Mon. Not. R. Astron. Soc.* **434** 1565–1575
[4] Cooray A and Chen X 2002 *Astrophys. J.* **573** 43–50
[5] Chluba J and Mannheim K 2002 *Astron. Astrophys.* **396** 419–427
[6] Sunyaev R A, Norman M L and Bryan G L 2003 *AstL* **29** 783–790
[7] Hwang H S and Lee M G 2007 *Astrophys. J.* **662** 236–249
[8] Tovmassian H M 2015 *Astrophysics* **58** 328–337
[9] Manolopoulou M and Plionis M 2017 *Mon. Not. R. Astron. Soc.* **465** 2616–2633
[10] Sunyaev R A and Zel'dovich Y B 1970 *Astrophys. Space Sci.* **7** 20–30
[11] Sunyaev R A and Zel'dovich Y B 1980 *Annu. Rev. Astron. Astrophys.* **18** 537–560
[12] Sembolini F *et al.* 2013 *Mon. Not. R. Astron. Soc.* **429** 323–345
[13] Baldi A S *et al.* 2017 *Mon. Not. R. Astron. Soc.* **465** 2584–2594
[14] Prada F *et al.* 2012 *Mon. Not. R. Astron. Soc.* **423** 3018–3030
[15] Mather J C *et al.* 1999 *Astrophys. J.* **512** 511–520
[16] Vikhlinin A *et al.* 2006 *Astrophys. J.* **640** 691–709

Evaluating a proxy of the local entropy production rate on the solar photosphere

G Viavattene[1], F Berrilli[1], G Consolini[2], D Del Moro[1], F Giannattasio[3], L Giovannelli[1], V Penza[1]

[1] Università degli Studi di Roma "Tor Vergata", Via della Ricerca Scientifica, 1, 00133 Rome, Italy

[2] INAF - Istituto di Astrofisica e Planetologia Spaziali (INAF-IAPS), Via del Fosso del Cavaliere, 100, 00133 Rome, Italy

[3] Istituto Nazionale di Geofisica e Vulcanologia, Via di Vigna Murata, 605, 00143 Rome, Italy

E-mail: giorgio.viavattene@roma2.infn.it

Abstract. The evaluation of the entropy production rate on the solar photosphere and its probability distribution are the key issues for studying the non-equilibrium dynamics of the solar convection. The local entropy production rate can be easily evaluated using the vertical heat flux as a proxy, which is given by a product between the line-of-sight velocity and the surface temperature. In this framework, the solar photosphere provides an incomparable laboratory to study turbulent convection in a dissipative non-equilibrium system near a steady state. In this work, we present some preliminary results on statistics of the local entropy production rate via the vertical heat flux, using line-of-sight velocity and temperature maps of the solar photosphere which are extracted from spectro-polarimetric data making use of the Center of Gravity Method and the simple black body radiation law.

1. Introduction

Most of the natural systems exist in a non-equilibrium configuration, showing very intriguing dynamics. A thermodynamic description of such systems requires the investigation of the entropy production rate [1], which is the physical parameter that quantifies the non-equilibrium state of these systems.

Among the uncountable non-equilibrium systems, the Sun is one of the most interesting, representing a natural laboratory to investigate the nature and the physics of turbulent convection with high Rayleigh number [2–4]. Nowadays, space and ground-based observations enable us to investigate this physical problem in an unprecedented detail, providing high-resolution images of several physical quantities at the Sun's surface. This allow us to investigate in a great detail the dynamics of solar convection on the solar photosphere.

In non-equilibrium systems the measure of the entropy production rate is one of the key problems of non-equilibrium statistical thermodynamics, which presents many unanswered questions. Linear and non-linear non-equilibrium thermodynamics predicts that in non-equilibrium systems there is a non-zero spontaneous entropy production rate, which, according to the second law of thermodynamics, is a manifestation of the irreversible nature of such systems [1]. In non-equilibrium systems the entropy production rate is a function of position $\mathbf{r}$ and time t, Which can be written as:

$$\sigma(\vec{r},t) = \sum_i J_i(\vec{r},t) X_i(\vec{r},t) > 0 \tag{1}$$

where $J_i(\vec{r},t)$ is a generic thermodynamic flux quantity and $X_i(\vec{r},t)$ is the associated generalized thermodynamic force or affinity, and the i index accounts for the different contributions to the entropy production rate. In treating the solar photospheric convection, we will concentrate on the vertical heat flux, $\vec{J}_q(\vec{r},t)$, which will be estimated on the basis of the line-of-sight (LoS) velocity and the solar surface temperature.
Here, we present a preliminary evaluation of the statistical features of the entropy production rate in the solar photospheric layer, via the heat flux. Similar studies have been carried out in experimental laboratory physics. The statistical properties of the heat flux has been studied in a cylindrical cell filled with water in [5], and in a von Karman experimental setup and in a wind tunnel in [6].

2. The turbulent solar convection

The solar convection is a process that occurs in an internal layer of the Sun, the so-called convection zone (CZ). In this region, which is unstable according to the Schwarzschild criterion [7], a large fraction of the energy is carried by turbulent plasma flows, with the hotter plasma ascending and the colder plasma descending, driven by the surface entropy sink (radiative cooling). The signature of the convective pattern at the main energy carrying (granular) scale is visible in the solar photosphere as a collection of bright granules, with a typical size of 1 Mm, surrounded by a network of dark intergranular lanes [7–10]. If we define as CZ upper boundary the surface where the radial component of the upward velocity of the convective flows goes to zero [11], the CZ includes what is called the photospheric overshooting region.
The turbulent solar convection can be viewed as a dissipative process near a non-equilibrium steady state (NESS) and the Sun provides an unequaled laboratory to perform the analysis of the entropy production rate in these systems.
The analysis shown in this work is performed by using the observations of a quiet-Sun region, i.e., a region excluding intense magnetic patches, in which we can assume that the convection pattern is not altered by the presence of magnetic field. In such an approximation, we can neglect the back-reaction of the magnetic field on the plasma motion, i.e., $\vec{J} \times \vec{B} \simeq 0$, and thus, the equations governing the solar convection in such quiet regions reduce to the standard fluid equations for heat-transport, i.e.,

$$\partial_t \vec{v} + (\vec{v} \cdot \nabla)\, \vec{v} = -\frac{1}{\rho} \nabla p + \nu \nabla^2 \vec{v} - \nabla \Omega \tag{2}$$

$$\partial_t T + \vec{v} \cdot \nabla T = \kappa \nabla^2 T \tag{3}$$

where Ω is the gravitational potential, ν is the plasma viscosity, κ is the temperature diffusivity coefficient, $\vec{v}$ the velocity, T the temperature and p the pressure.
In this regime, the entropy production rate can be written in terms of the *heat flux*, $\vec{J}_q$, and the temperature gradient, ∇T, as follows [1],

$$\sigma = -\frac{1}{T^2} \vec{J}_q \cdot \nabla T, \tag{4}$$

so that assuming the temperature gradient to be constant, $\nabla T = const$, the heat flux provides a proxy of the entropy production rate. The last equation can be written also in the following form,

$$\sigma = \vec{J_q} \cdot \nabla \frac{1}{T}, \tag{5}$$

which will be used in the next Section to estimate the entropy production rate σ (see also [5] and references therein).

3. The evaluation of a proxy of the entropy production rate

The solar surface convection is a good example of phenomenon in a NESS. Thus, on the basis of the previous discussion and following [5], we can evaluate the local entropy production rate $\sigma(\mathbf{r}, t)$ directly from the vertical heat flux $j_z(\mathbf{r}, t)$, assuming that the transport occurs mainly in the vertical (radial) direction:

$$\sigma(\vec{r}, t) \approx V_0 j_z(\vec{r}, t) \nabla_z \left(\frac{1}{T}\right) \tag{6}$$

where V_0 is the volume over which the local properties are evaluated and $\nabla_z \left(\frac{1}{T}\right)$ is the vertical gradient of the temperature T which is responsible for maintaining the convection.
Therefore, the local vertical heat flux is used as a *proxy* of the local entropy production rate and it can be computed as follows:

$$j_z(\vec{r}, t) \approx v_{LoS}(\vec{r}, t) \delta T(\vec{r}, t) \tag{7}$$

where $v_{LoS}(\vec{r}, t)$ is the plasma velocity along the line-of-sight (LoS) and $\delta T(\vec{r}, t) = T(\vec{r}, t) - T_0$, with T_0 being the average bulk temperature.

4. Dataset and data analysis

The dataset used for this analysis has been acquired on November 21st 2006 with the Interferometric BIdimensional Spectropolarimeter (IBIS [12, 13]) installed at the Dunn Solar Telescope (DST), located in National Solar Observatory (NSO), Sacramento Peak, New Mexico. IBIS can perform imaging spectral scans of a selection of spectral regions with a resolution of $\simeq$ 4 pm, and in its spectropolarimetry mode, it can also measure the polarization state of the incoming light, thus acquiring the Stokes profiles (the total intensity I, the excess of linear polarization Q and U, and the excess of circular polarization V) of the spectral lines. For this dataset IBIS measured the four Stokes profiles in the spectral region containing the Fe I 630.15 nm, and 630.25 nm, lines and broadband images in a nearby spectral region. The Field-of-View (FoV) of data is at the solar disk center, so in our dataset the radial direction coincide with the line-of-sight (LoS) direction. The FoV imaged by the instrument is 40×40 arcsec^2, which corresponds approximately on the solar surface to 30×30 Mm^2. The spatial resolution is 0.17 arcsec, corresponding to $\simeq$120 km on the solar surface, the time resolution is 89 seconds (which is the time to perform a complete spectral scan) and the whole duration of the dataset is about one hour (41 time steps in total). For more details and information on the dataset, see [14–17]. The dataset has been calibrated using the IBIS pipeline [18]. In Fig. 1 (left panel) the spectral scanning of Stokes I parameter performed by the IBIS instrument is shown. Both spectral lines of interest are clearly visible.
The entropy production rate has been calculated from temperature and velocity maps, using Eq. 7. In detail, the temperature maps are evaluated using the black body radiation law [19, 20] applied to the broadband images. In particular, we computed the temperature maps from the Stefan-Boltzmann law as follows:

$$T(\vec{r},t) = T_{eff}\sqrt[4]{C(\vec{r},t)} \tag{8}$$

where $T_{eff} = 5780$ K is the average temperature of the solar photosphere [7] and the contrast has been defined as:

$$C(\vec{r},t) = I(\vec{r},t)/\bar{I}(t) \tag{9}$$

with $\bar{I}(t)$ being the average of the intensity I for each temporal frame.

We evaluated the LoS velocity maps using the Center of Gravity (CoG [21–23]) Method applied to the Fe I 630.15 nm spectral line, also discussed and compared with inversion techniques in [17], as follows:

$$v_{LoS}(\vec{r},t) = \frac{\int I(\vec{r},t)\lambda d\lambda}{\int I(\vec{r},t)d\lambda} \tag{10}$$

where λ the wavelength and the integral is extended over all the wavelengths acquired during the spectral scan.

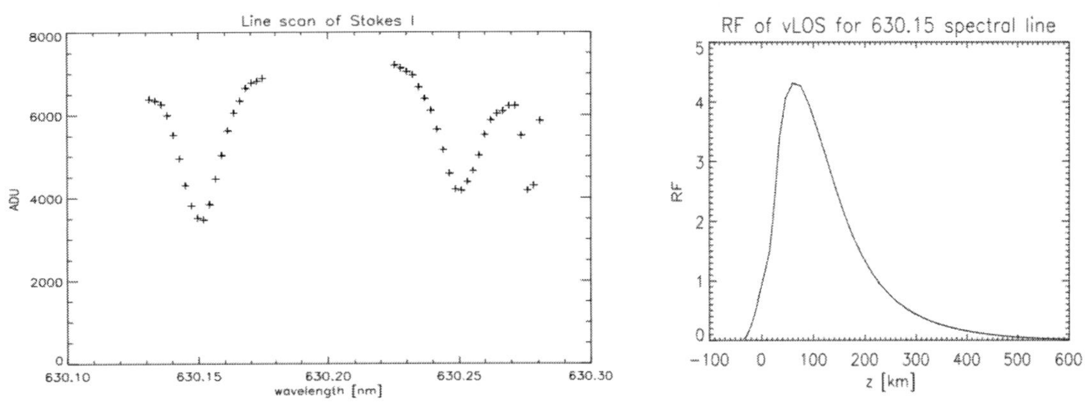

Figure 1. Left panel: Average IBIS scan of Stokes I profile of Fe I 630.15 nm and 630.25 nm spectral lines. At 630.28 nm is visible a telluric line due to absorption of water in the Earth's atmosphere. Right panel: average RF to v_{LoS} for Fe I 630.15 nm, spectral line.

Since we are using the Stefan-Boltzmann law applied to broadband images to compute the temperature maps, we assume that such temperature is associated with the base layer of the photosphere within the photon mean free path ($\approx$ 100 km [24]). The association of a specific solar atmosphere layer with the LoS velocity maps instead deserves a more detailed treatment. To estimate the height of the v_{LoS} signal we are measuring, we make use of Response Functions (RFs) [25–27]. RFs tell us which is the atmospheric layer where the Stokes profiles are sensible to perturbation in the atmospheric parameters (such as temperature, LoS velocity, magnetic field intensity, and so on), and so the atmospheric height where the information encoded in the spectral line is mainly formed. Using standard solar atmosphere parameters, we are able to evaluate the RFs to v_{LoS} for each measured wavelength of the Fe I 630.15 nm spectral line [26]. By averaging those RFs, we obtain the RF to v_{LoS} tailored to our measure, shown in Fig. 1 (right panel). From the RF shape, using the full-width half-maximum, we can estimate that the layer associated with our v_{LoS} maps is $\approx 70^{+80}_{-50}$ km above the base of the solar photosphere. Therefore, the T and v_{LoS} signals we have measured, are associated to the same layer within

the measured errors, and we can safely combine them to compute a vertical heat flux J that can be associated with the same layer in the solar atmosphere.
As customarily done in solar convention studies, the temperature and LoS velocity maps have been filtered through a subsonic $k_h - \omega$ filter, in order to remove the signal from the acoustic oscillations in the solar photosphere [28].

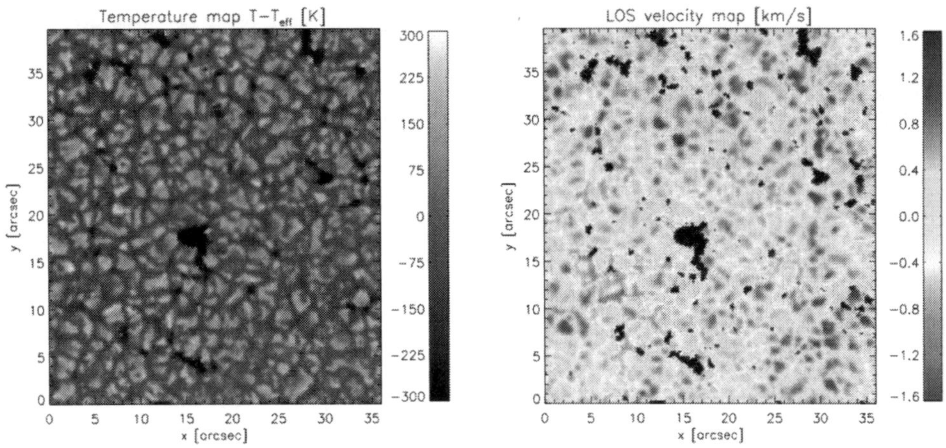

Figure 2. Left panel: a sample of a temperature map evaluated using the black body radiation law. The map shows the temperature fluctuations from the average temperature of the solar photosphere (T_{eff}). Right panel: the co-temporal LoS velocity map evaluated using the CoG method. Black pixels mask those regions excluded from our analysis because they have magnetic field intensity greater than 50 Gauss.

Using the polarimetric signal in the Stokes V parameter, we masked out the region with an estimate magnetic flux greater than ≈ 50 Gauss (black regions in Figure 2 and left panel of 3), in order to analyze a quiet, i.e., non-magnetic, region.

An example of the map of the vertical heat flux J, evaluated with Eq. 7 is reported in Fig. 3 (left panel). The histogram of the vertical heat flux values for all the time steps is reported in Fig. 3 (Right panel). The distribution of J is visibly asymmetric. This confirms that on the solar photosphere there is a spontaneous and positive production of entropy, typical of non-equilibrium systems.

5. Conclusion

In this work we investigate the entropy production rate on the solar photosphere using the vertical heat flux as a proxy. The latter was evaluated using temperature and LoS velocity maps retrieved from broadband images and spectro-polarimetric data using the black body radiation law and the CoG method, respectively.
We show that the observed distribution of the vertical heat flux is asymmetric and this suggests that the solar convection represents a natural non-equilibrium system, characterized by a non-zero entropy production rate.
The preliminary results we obtained pave the way towards future analyses reaching a complete characterization of the non-equilibrium dynamical state of solar photospheric convection.

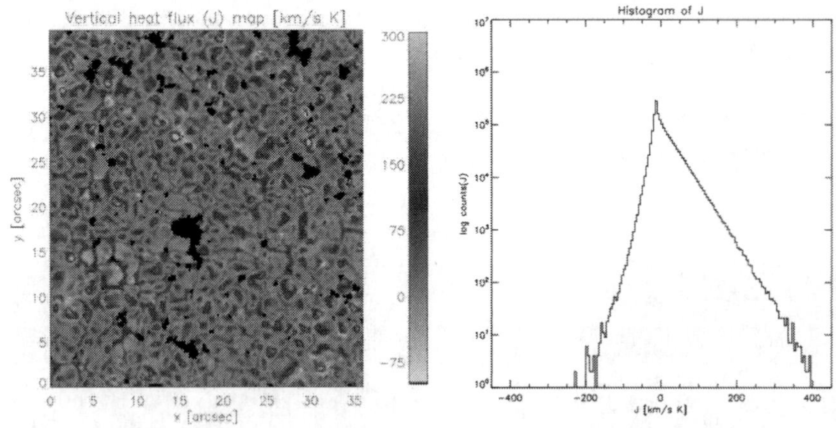

Figure 3. Left panel: sample vertical heat flux map. As above, black pixels mask those regions excluded from analysis. Right panel: histogram of the vertical heat flux for the whole dataset.

References

[1] de Groot SR and Mazur P 1994 *Non-equilibrium Thermodynamics* Dover Pub. Inc., New York

[2] Berrilli F, Del Moro D, Russo S, Consolini G, and Straus T 2005 *Astrophysical Journal* **632** 677

[3] Giannattasio F, Del Moro D, Berrilli F, Bellot Rubio L, Gosic M, and Orozco Suarez D 2013 *Astrophysical Journal* **770** L36

[4] Caroli A, Giannattasio F, Fanfoni M, Del Moro D, Consolini G, and Berrilli F 2015 *Journal of Plasma Physics* **81(5)** 495810514 doi:10.1017/S0022377815000872

[5] Shang X D, Tong P and Xia K Q 2005 *Physical Review E* **72** 015301(R)

[6] Ciliberto S, Garnier N, Hernandez S, Lacpatia C, Pinton J F and Ruiz Chavarria G 2004 *Physica A* **340** 240-250

[7] Landi Degl'Innocenti E 2008 *Fisica Solare* **Springer**

[8] Berrilli F, Consolini G, Pietropaolo E, Caccin B, Penza V and Lepreti F 2002 *Astronomy & Astrophysics* **381** 253

[9] Del Moro D 2004 *Astronomy & Astrophysics* **420** 1007

[10] Nordlund Å, Stein R F and Asplund M 2009 *Living Reviews in Solar Physics* **6**:2 https://doi.org/10.12942/lrsp-2009-2

[11] Roxburgh, I. W. 1978, *Astronomy & Astrophysics*, **65**, 281

[12] Cavallini F, Berrilli F, Cantarano S and Egidi A 2001 *Mem. SAIt* **72** 554

[13] Cavallini F 2006 *Solar Physics* **236** 415

[14] Viticchié B, Del Moro D, Berrilli F, Bellot Rubio L and Tritschler A 2009 *Astrophysical Journal* **700** L145

[15] Del Moro D, Giannattasio F, Berrilli F, Consolini G, Lepreti F and Gošić M 2015 *Astronomy & Astrophysics* **576** A47

[16] Del Moro D, Giovannelli L, Pietropaolo E and Berrilli F 2017 *Experimental Astronomy* **43** 23

[17] Viavattene G, Berrilli F, Collados Vera M, Del Moro D, Giovannelli L, Ruiz Cobo B and Zuccarello F 2018 *IOP Conferences Series: Journal of Physics: Conf. Series* **956** 012006

[18] Viticchié B, Del Moro D, Criscuoli S and Berrilli F 2010 *The Astrophysical Journal* **273** 787

[19] Rodríguez Hidalgo I, Collados M and Vázquez M 1992 *Astronomy & Astrophysics* **264** 661

[20] Caccin B and Penza V 2000 *Proceeding of the 1st Solar and Space Weather Euroconference ESA SP* **463** 680

[21] Semel M D 1967 *Annales d'Astrophysique* **30** 513

[22] Rees D E and Semel M D 1979 *Astronomy & Astrophysics* **74** 1R

[23] Landi Degl'Innocenti E and Landolfi M 2004 *Astrophysics and Space Science Library* **307** L

[24] Judge P G, Kleint L, Uitenbroek H, Rempel M, Suematsu Y and Tsuneta M 2015 *Sol. & Phys.* **290** 979

[25] Penza V, Caccin B and Del Moro D 2004 *Astronomy & Astrophysics* **427** 345

[26] Caccin B, Gomez M T, Marmolino C and Severino G 1977 *Astronomy & Astrophysics* **54** 227

[27] Penza V, Caccin B, Ermolli I and Centrone M 2004 *Astronomy & Astrophysics* **413** 1115

[28] Title A M, Tarbell T D, Topka K P, Ferguson F H, Shine R A and the SOUP Team 1989 *The Astrophysical Journal* **336** 475

IOP Conf. Series: Journal of Physics: Conf. Series **1226** (2019) 012005 doi:10.1088/1742-6596/1226/1/012005

Optical analysis of protein–metal interactions

M A Baranov, A I Dudina and E K Nepomnyaschaya

Peter the Great St.Petersburg Polytechnic University (SPbPU) St.Petersburg, Polytechnicheskaya, 29, 195251, Russia

E-mail: baranovma1993@gmail.com

Abstract. In this paper we considered the processes of interaction of albumin protein molecules and ferrofluids. Our experimental studies allowed measuring the size of the protein molecules bound to ferrofluid particles. In addition, the structures ша the self-assembled protein films were obtained. The dependence of the structures formation from the magnetic liquids concentration in the protein solutions is shown. The findings may be useful for medical applications.

1. Introduction

Today new methods of therapy of the different diseases are rapidly developing. Drug delivery to the lesion in the body, in particular to tumors, is one of the main directions of explorations [1]. It has been proposed that ferrofluid can be applied for the target drug delivery. However, this technique is still under development. Ferrofluid is a special category of smart metallic nanomaterials in colloidal suspensions composed of single-domain magnetic nanoparticles [2]. Possible methods of the application of ferrofluids in medicine were considered in previous studies [3].

In this study the interaction between albumin protein molecules and ferrofluid particles is considered. Albumin molecule has lots of binding sites [4]. The particles of ferrofluid can settle on the binding sites and disturb transport properties of albumin. If this function is violated, the whole body will suffer. Therefore, it is necessary to investigate albumin transport properties in the presence of a magnetic fluid. Additionally, self-organized albumin films with particles of ferrofluid were studied. It is experimentally shown that the processes of protein self-organization, including albumin, can occur with the dehydration of their water solutions. It was established that drying the protein film results in the formation of a specific ordered morphological structure [5–8]. The research of the protein–metal self-organization is necessary for the understanding the protein functioning.

In this study, we aim to study and observe the interaction albumin molecules with particles of ferrofluid.

2. Methods

In this work, the method of depolarized dynamic light scattering (DDLS) was used. This method allowed us to determine the sizes of small particles (including non-spherical particles) in liquid. DDLS is based on the measurement of translation and rotational dynamic coefficients of particles [9]. With these coefficients, the shape and size of particles can be calculated.

DDLS method's theory is based on the theory of Rayleigh scattering of the coherent light [10]. This theory works for particles with dimensions d much smaller than the wavelength of the incident light λ ($d \sim 0.03\lambda$). In this condition the elastic scattering occurs, the scattered radiation remains coherent, and the wavelength remains constant $\lambda' = \lambda$. If the motion of the considered particles is Brownian, which is the case in absence of external influences, then quasi-elastic scattering ($\lambda' \approx \lambda$) will occur. However this shift is too small in comparison with the wavelength of the incident light, so more often the theory of speckle fields' dynamics is used [11].

Because the coherent incident light is scattered by particles, which perform the Brownian motion, the dynamic speckle field will be formed at the detection plane. With each shift of the particles, a new realizations of the speckle field will occur. The change in the arrangement of the scatterers is the reason for the difference in the realizations of the speckle field [12]. This is called speckle dynamics.

If we measure the time dependence of scattered light intensity fluctuations (integrated in the small detection area), we will be able to calculate the coefficients of translational and rotational diffusion and also the sizes of the particles under study [13].

The theory of DDLS was described in our previous work [14]. We explained how from simple scattering equations for polarized (VV) and depolarized (VH) components of scattered light we can calculate two time-dependent autocorrelation functions (here, we present only some of the simplest equations that hold for particles with rotational symmetry such as ellipsoids and cylinders).

$$G^{(VV)}(\tau) = S_1(q,R)e^{-q^2 D_T \tau} + S_2(q,R)e^{-(q^2 D_T + 6D_r)\tau} = S_1(q,R)e^{-\Gamma_1 \tau} + S_2(q,R)e^{-\Gamma_2 \tau} \quad (1)$$

$$G^{(VH)}(\tau) = S_3(q,R)e^{-(q^2 D_T + 6D_r)\tau} = S_3(q,R)e^{-\Gamma_2 \tau}$$

where q=(4πn/λ)sin (θ/2) is the scattering wave vector, n is the refractive index of the medium, λ is the wavelength of the incident radiation, θ is the angle of the scattering detection, D_T is the translational diffusion coefficient, and D_R is the rotational diffusion coefficient.

The value of diffusion coefficients depends on the shape of the scatterers. In particular, so for ellipsoids of rotation with the semi axes R_a and R_b we can give the following equations

$$D_T = \frac{k_b T}{6\pi\eta R_a} F(R_a, R_b),$$

$$D_R = \frac{k_b T}{6\pi\eta R_a} \frac{\left(2 - \frac{{R_b}^2}{R_a}\right) F(R_a, R_b) - 1}{1 - \frac{R_b^4}{R_a}}, \quad (2)$$

$$F(R_a, R_b) = \frac{1}{\sqrt{1 - \left(\frac{R_b}{R_a}\right)^2}} \cdot \ln\left(\frac{1 + \sqrt{1 - \left(\frac{R_b}{R_a}\right)^2}}{\frac{R_b}{R_a}}\right),$$

where η is the viscosity of the solvent, k_b is the Boltzmann constant, T is the temperature, $R_a < R_b$.

We used the DDLS setup (Fig. 1) to measure the autocorrelation functions. The light from a semiconductor laser source was polarized and focused in a cuvette containing the sample solution. The laser irradiation wavelength was selected depending on the absorption of the sample. In this case we selected λ=655 nm [15]. The scattered light intensity was registered at 90° angle by a photomultiplier. The detect to state of polarization of the scattered light we used a second polarizer placed before the photomultiplier. The detected signal was digitized at the 100 kHz frequency and processed with a computer, as described in [14].

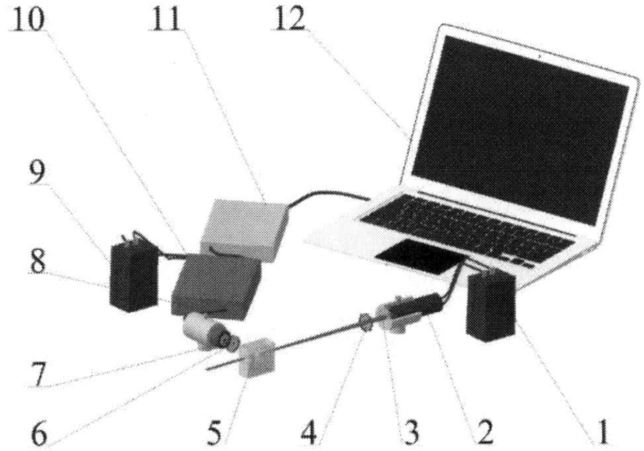

Figure 1. Depolarized dynamic light scattering device. 1 — accumulator battery, 2 — semiconductor laser, 3 — lens, 4 — polarizer, 5 — cuvette with analyzed sample, 6 — rotary polarizer, 7 — diaphragm, 8 — photomultiplier, 9 — power supply, 10 — analog-to-digital convertor, 12 — computer

The DDLS technique was used to show the ability of ferrofluid nanoparticles to interact with binding sites of albumin molecules. With these sites, the molecule usually transfers microelements through the human body. We decided to check if there are any influences of ferrofluids on the processes of albumin self-organization. This property of proteins is very important because of protein functionality, its transport properties, its ability of interaction with other molecules of proteins, ferments etc. [16].
Also in this study, the research of the self-assembled protein films with the ferrofluid was carried out. Water solution of albumin was prepared from concentrated protein albumin solution with pH = 9.3 and distilled water with pH = 7.0 in the ratio of 1:1, 1:2, 1:3 with ferrofluid in volume 1–3 μl. 2 ml of solutions were placed on a Petri dish. The solution dries out within 48 hours. The film's photos were obtained using the setup in Fig. 2:

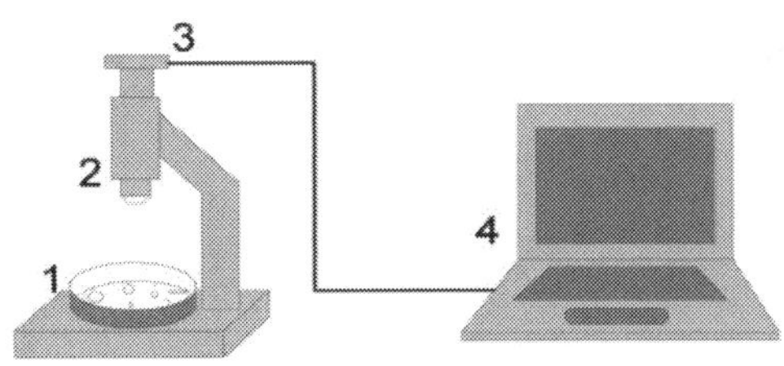

Figure 2. Experimental setup: 1 — cuvette with a sample, 2 — microscope, 3 — CCD camera, 4 — computer

3. Results and discussion

We characterized the solutions of the ferrofluid Fe_3O_4 with albumin protein by measuring the autocorrelation functions, and the diffusion coefficients using our experimental setup, and by calculating the aggregates' sizes in the ellipsoidal approximation. From the measured values of D_T and D_R, the sizes of the particles were calculated by the mathematical processing of the experimental data.
Distributions of Γ_1 and Γ_2 for polarized and depolarized components of scattered light for a solution of ferrofluid are shown in Fig. 1–2. The concentration of nanoparticles in ferrofluid was reduced by dilution with water till 0.01 vol. % to reproduce the single-scattering condition.

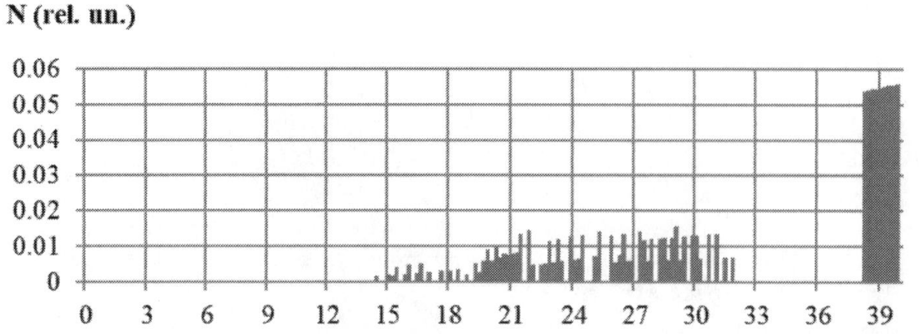

Figure 3. Distribution of relative concentrations of ferrofluid nanoparticles by Γ for a polarized component of scattered light

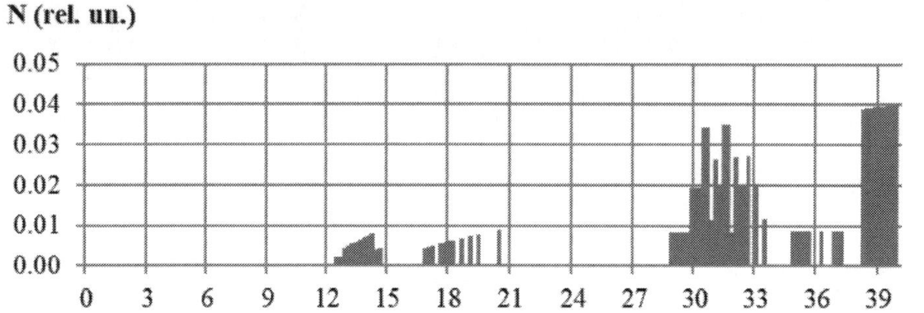

Figure 4. Distribution of relative concentrations of ferrofluid nanoparticles by Γ for a depolarized component of scattered light

The same type of measurements was carried out for albumin water solution with a diluted concentration of 10 mg/ml and the mixture of albumin and ferrofluid solutions. In the following table, the resulting sizes and diffusion coefficients for nanoparticles in the described mixtures are presented.

Table 1. Sizes of particles and their diffusion coefficients in the studied mixtures

Sample	$D_T \cdot 10^{-12}$ (m²/s)	D_R (1/s)	R_a (nm)	R_b (nm)
Magnetic fluid	6.23	6254–5368	22–25	11–12
Albumin	9.6	16178	19	6
Mixture	4.42	1599–2170	32–41	13–17

As it is shown in Tab. I, the sizes of particles in albumin and ferrofluid mixture are larger than in the initial solutions.

Experimental solutions from the previous experiment were dehydrated. The structures in the obtained films were studied by the experimental setup in Fig. 2. Experimental results are shown in Fig. 5.

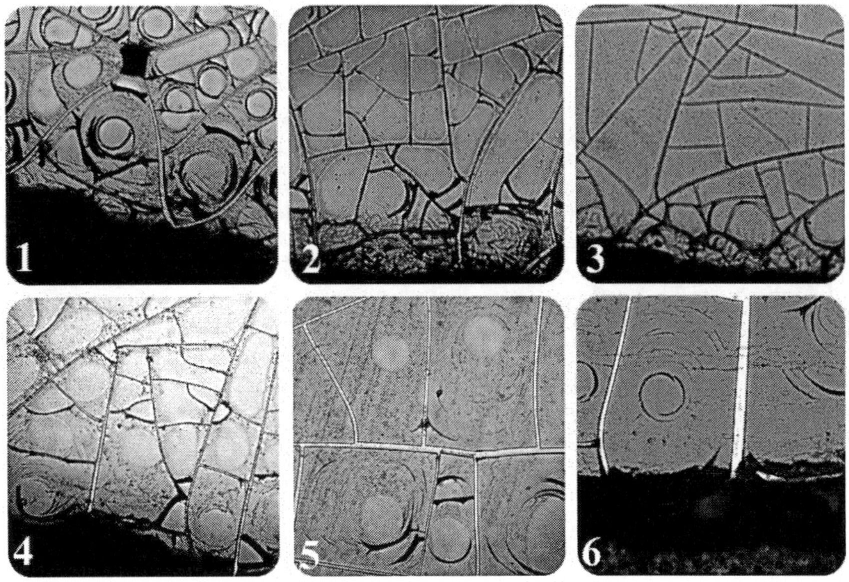

Figure 5. The photos of albumin films: 1 – water albumin solution 1:1 (protein: water), 2 - water albumin solution 1:2, 3 – water albumin solution 1:3, 4 - water albumin solution 1:1 with 1 μl of ferrofluid, 5 – 2 μl of ferrofluid, 6 – 3 μl of ferrofluid

Fig. 5 shows that the concentration of protein in water solution influences the structures formed in the films. Spiral structures were distracted at high water concentration. Addition of the ferrofluid into the albumin solution affects the films structures. Self-organization is violated because the communication between albumin molecules is hindered by ferrofluid particles.

4. Conclusion

We studied the interaction of albumin molecules with ferrofluid particles. Ferrofluid particles settle on the binding sites of the protein molecules. Interestingly, we showed the formation of agglomerates, whose sizes are larger than the albumin molecules and the ferrofluid particles alone. Additionally, we demonstrate that the spiral structures in the protein film depend on the concentration of the albumin molecules and the ferrofluid particles in the protein solutions.

In summary, our results confirm the capability of ferrofluid particles to bind organic molecules, opening the possibility of using ferrofluid particles for drug transportation.

References

[1] Latorre M, Rinaldi C 2009 *Puerto Rico health sciences journal* **28** 227
[2] Scherer C, Figueiredo Neto A M 2005 *Brazilian Journal of Physics* **35** 718
[3] Nepomnyashchaya E K, Velichko E N, Pleshakov I V, Aksenov E T, Savchenko E A 2017 *Journal of Physics: Conference Series* **841**
[4] Andersen J T, Dalhus B, Cameron J, Daba M B, Plumridge A, Evans L, Sandlie I. 2012 *Nature communications* **3** 610
[5] Velichko E, Baranov M, Nepomnyashchaya E, Cheremiskina, A, Aksenov E 2015 *Conference on Smart Spaces* 786
[6] Baranov M A, Velichko, E N, Aksenov E T 2017 *Journal of Physics: Conference Series* **917**
[7] Baranov M A, Velichko E N, Nepomnyashchaya E K, Aksenov E T 2017 *Progress In Electromagnetics Research Symposium-Spring (*IEEE*)* 42
[8] Rajagopal K, Schneider J P 2004 *Current opinion in structural biology* **14** 480
[9] Lehner D, Lindner H, Glatter O, 2000 *Langmuir* **16** 1689
[10] Stetefeld J, Sean A 2016 *Biophys Rev.* 409

[11] Murphy R J, Pristinski D, Migler K, Douglas J F, Prabhu V M 2010 *The Journal of Chemical Physics* 132
[12] Stetefeld J, McKenna S A, Patel T R 2016 *Biophys Rev* 409
[13] Tereshchenko S A, Burnaevskiy I S, Dolgushin S A, Yudin I K, Deshabo V A 2017 *Biomedical Engineering* **50** 385
[14] Nepomnyashchaya E, Aksenov E, Velichko E 2017 *Spring PIERS* 3556
[15] Nepomnyashchaya E K, Prokofiev A V, Velichko E N , Pleshakov I V , Kuzmin Yu I 2017 *Journal of Magnetism and Magnetic Materials* **431** 24
[16] Wollman A J, Sanchez-Cano C, Carstairs H M, Cross R A, Turberfield A J 2014 *Nature nanotechnology* **9** 44

IOP Conf. Series: Journal of Physics: Conf. Series **1226** (2019) 012006 doi:10.1088/1742-6596/1226/1/012006

Radiochromic films in Radiation Hardness Space Application

P Casolaro, L Campajola, D De Luca

University of Napoli Federico II, Department of Physics, I-80126 Napoli, Italy

E-mail: casolaro@na.infn.it

Abstract. The characteristics of radiochromic films such as the direct visualization of radiation field, ease of use and data analysis are suited for the dosimetry monitoring in tests of radiation hardness space applications. In particular, in this work, in order to study the dependence of radiation type, energy and dose rate, as well as the dynamic range of EBT3 Gafchromic films, a set of films was exposed to radiation sources used in total ionizing dose (^{60}Co gamma- and $^{90}Sr/^{90}Y$ beta-rays). The results have been found to be particularly suited for the employment of this kind of films in radiation hardness assurance tests.

1. Introduction

The electronic devices employed in space applications (e.g. satellites, cubesats, spacecraft, batteries, sensors, power units, payload equipment, among others) need to be tested due to the high exposure to radiation in the space. The space radiation environment is made of several radiation types and energies, not homogeneously distributed in space and time. In particular, it is composed by particles trapped by Earth's magnetic field, Galactic Cosmic Rays (GCRs) and Solar Cosmic Rays (SCRs). The terrestrial magnetic field traps charged particles such as electrons and protons, by creating two radiation belts known as Van Allen belts [1]. The inner belt, which extends from 100 to 6000 km in altitude is known as Low Earth Orbit (LEO) and is mainly populated by high-energy protons in the range 10-100 MeV with density 15 particles/m^3 and electrons in the range 50-1000 keV. The outer belt is populated by electrons up to 10 MeV at altitudes up to 60000 km. Moreover, because of the difference between the magnetic and geographical poles, the Van Allen belts are closer to one part of Earth that is the South Atlantic area. The subsequent radiation anomaly is known as South Atlantic Anomaly (SAA). GCRs follow helical trajectories because of their interaction with strong magnetic fields in the galaxy; GCRs and SCRs mainly consist of protons with a lower contribution of alfa particles, but also heavy nuclei are part of primary cosmic rays. For these reasons, radiation tests for quality checks of electronics must be accurately designed by taking into account the typical characteristics of the space mission such as time, orbit, altitude and shielding.

The electronic devices employed in space missions must be therefore declared radiation hardened or tolerant before their usage. A useful schematization of the effects of radiation on silicon can be done in term of long- and short-term effects. The firsts are total ionizing and displacement damage effects and the seconds are Single-Event-Effects (SEEs). Total Ionizing Dose (TID) tests are usually performed with gamma rays from radioactive sources of ^{60}Co, but also protons from accelerators in the range from a few keV to 500 MeV or electrons in a wider range from some eV to 10 MeV. TID causes long-term degradation of electronic devices due to cumulative energy deposited in the Si-matrix for ionization processes. Displacement Damage Dose (DDD) also implies long-term degradation of electronic devices, but in this case the basic mechanism of damage consists in the displacement of the atoms of the Si-lattice. The characterization of devices for DDD is more complicated than that for TID. In DDD tests, the contribution to non-ionizing energy loss (NIEL) should be accomplished, in agreement with the formula

$$DDD = NIEL \cdot \Phi$$

where Φ is the particle fluence in cm^{-2}. Similarly, for TID tests, the evaluation of linear energy transfer (LET) has to be done accordingly to the formula

$$TID = LET \cdot \Phi$$

Tables of both LET and NIEL values are available in literature [2]. Fig. 1 shows a picture with a SEE on an integrated circuit. It results in the direct breakdown of the device.

Fig. 1: Picture of a SEE on an integrated circuit.

SEEs are stochastic and occur when a single ion interacts with the DUT Si-matrix, by inducing direct or indirect ionization (e.g. neutrons). SEE probability is a characteristic function of the LET of the incident ion.
The characterization of electronic devices for TID, DDD and SEE tests is usually performed in facilities able to provide charged particle beams and radioactive sources of suited characteristics. The guide-lines for the testing of devices, defined by the International Standards such as the U.S. military standard guidelines MIL-STD-883 and the ESA ESCC22900 standard guidelines [4,5], impose conditions for radiation hardness quality assurance tests and consequently for the typologies of dosimeters and radiation detectors employed in these tests. In this work, a full characterization of some types of radiochromic films, the EBT3 Gafchromic films, has been accomplished for the specific application of radiation hardness assurance tests.

2. Radiochromic Films

Nowadays several detectors and dosimeters are available and each detector has its own features which make it more suitable for some specific application rather than another. Detectors and diagnostic techniques typically employed in radiation hardness applications are e.g. the gas-filled chambers such as ionization chambers and parallel-plate avalanche counters, Faraday cup, resistive or capacitive pick-up systems, scintillators and phosphor screen, among the others. The required information is often achieved by making use of more than one detector. Furthermore, in tests with particle beams it is common that the energy needed can be provided by reducing the initial incident energy. This can be performed by means of suited materials (degraders) that slow down and straggle the particles. The change of characteristics of the original beam leads to the loss of definition of relevant physical quantities like energy, fluence and dose in the experimental position. As an example, Fig. 2 shows a picture of the beam line of the proton irradiation facility (PIF) at *Paul Scherrer Institute* (PSI).

Fig. 2: PSI PIF beam line for irradiation of electronic devices.

It is noticeable the series of copper degraders needed for the modulation of the initial energy. In order to accurately monitor physical quantities needed for guarantee the reliability of different types of tests (TID, DDD and SEE) in radiation hardness applications, the ideal detector should be independent of incident radiation type, energy and dose rate, as well as non-perturbative and easy to use. A good compromise between these needs is the Radiochromic Film (RCF).

RCFs consist of a thin film made of a single or double layer of radiation-sensitive material that has the chemical property to exhibit an increased darkness with the increase of the radiation dose [6]. RCFs are commonly used in photon and ion radiotherapy and in many other applications of radiation physics such as industrial irradiations for modification of materials, food and medical instrumentation sterilization, among the others. The main characteristics of RCFs are reliability, ease of use, cost, and portability. Furthermore, RCFs are non-invasive dosimeters providing accurate and permanent values of dose. The direct visualization of the radiation field and micrometric thickness of the active layers are very useful characteristics in radiation hardness applications. The interaction of radiation with the radiochromic medium causes a progressive darkening of the film that can be related to the radiation dose. Differently from silver-halide films, RCFs are self-developing and the reading, usually made with commercial flat-bed scanners and densitometers, can be made after the time needed for the stabilization of the polymerization process which is taken for convenience 24 hour. Scanner and densitometers allow the evaluation of the Pixel Value (PV) on a particular region of interest of the digitized film. In this way the calibration (PV vs. dose) can be carried out by exposing a set of films to a-priori known radiation field. Several types of RCFs are currently available and a wide dynamic range can be investigated, from a few mGy up to several hundred of kGy. This range covers the most of doses in radiation hardness tests.

3. Results

In this Sec. we report the results of calibration of a set of EBT3 Gafchromic films, a kind of RCF intensively employed in medical physics application, to ^{60}Co gamma- and ^{90}Sr/^{90}Y beta-rays. It has to be noticed that ^{60}Co gamma sources constitute the standard sources for TID tests. The employment of a ^{90}Sr/^{90}Y beta-source for TID tests is a novelty in this field and represents a valid alternative to ^{60}Co sources [7]. In particular, the exposure of devices to beta-rays results in a better representation of the complex space radiation environment composed of several types, energies and dose-rates. The contribution of electrons to space radiation is mainly due to trapped electrons in the outer Van Allen belt. In addition, the use of a compact and easy handling beta source allows exposures of electronic devices without any damage to other auxiliary circuits. Fig. 3 shows the calibration (PV as a function of the dose) of a set of EBT3 Gafchromic films exposed to ^{60}Co gamma-rays from the GammaCell220 of ISOF-CNR in Bologna.

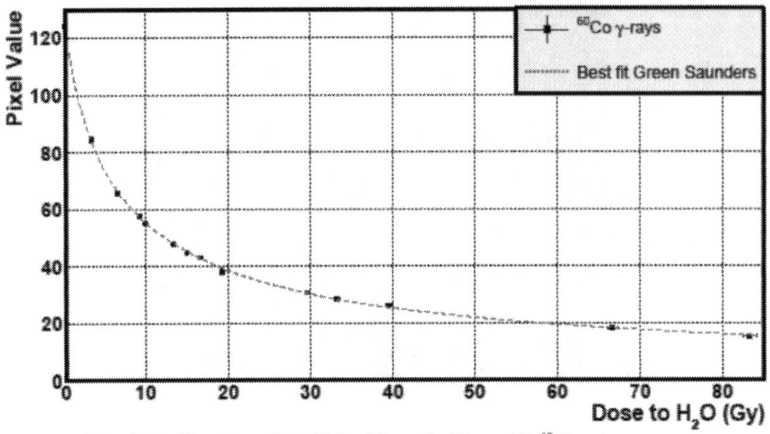

Fig. 3: Calibration of EBT3 Gafchromic films with ^{60}Co gamma rays.

The sensitivity range declared by the manufacturer is up to 40 Gy. For higher doses the sensitivity decreases, but it is possible to discriminate doses up 80 Gy (8 krad). The function which best fits to the experimental data has been found to be the Green-Saunders equation [8]. It is represented by the dotted red curve in Fig. 3. The agreement of the fit is excellent and the calculated value of the reduced chi-squared is 0.8. Fig. 4 shows the calibration of a set of EBT3 Gafchromic films to beta-rays from the radioactive source ^{90}Sr/^{90}Y.

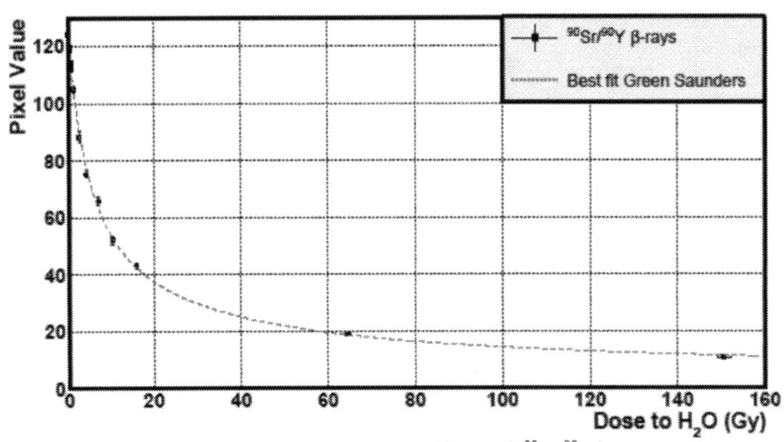

Fig. 4: Calibration of EBT3 Gafchromic films with ^{90}Sr/^{90}Y beta rays.

This source is currently used at University of Naples Federico II as accurate and reliable radioactive source for TID tests as alternative to the most common ^{60}Co. Also in this case, the best fit to the experimental data is provided by the Green-Saunders equation and the calculated value of the reduced chi-squared is 0.7. Data were acquired up to 160 Gy (16 krad).

4. Discussion and Conclusion

IOP Conf. Series: Journal of Physics: Conf. Series **1226** (2019) 012006 doi:10.1088/1742-6596/1226/1/012006

Tests of electronic devices in radiation hardness space application need reliable and easy to use instrumentation for the monitoring of dose and radiation field characteristics. Owing to the peculiar features of spatial radiation environment, consisting of several kinds of radiation, energies and dose-rates, radiochromic films are adapt to these purposes. In this work, a set of EBT3 Gafchromic films was exposed to the ^{60}Co gamma source (the mostly used one) and to the recent proposed ^{90}Sr/^{90}Y beta source. The results show that EBT3 Gafchromic films can be employed for most of tests carried out within this application. In fact, the dynamic range of these films goes from 1 cGy (1 rad) to hundreds of Gy (dozens of krad). For higher doses, other films, such as HDV2 Gafchromic films, sensitive in this range can be used.

References

[1] Campajola L and Di Capua F 2017 *Applications of Radiation Chemistry in the Fields of Industry, Biotechnology and Environment* Springer, Cham

[2] Messenger S R. *et al* 1999 *IEEE Transactions on Nuclear Science* **46.6** 1595-1602

[3] Berger M Z 1999 Stopping-Power and Range Tables for Electrons, Protons, and Helium Ions *Physical Reference Data*

[4] MIL-STD-883E Test Method 1019.4, Military Standards Test Methods and Procedures for Microelectronics, "Ionizing Radiation (total dose) Test Procedures", Defense Electronics Supply Center, December 2006

[5] European Space Agency, ECSS 22900, Total Dose Steady-State Irradiation Test Method. Online:http://escies.org/escc-specs/published/22900.pdf

[6] Soares CG 2006 Radiochromic film dosimetry *Radiation measurements* **41** S100-S116.

[7] Di Capua F *et al* 2018 A spot-based evaluation of TID performances for COTS components. in preparation

[8] Campajola L, Casolaro P and Di Capua F 2017 Absolute dose calibration of EBT3 Gafchromic films *Journal of Instrumentation* **12.08** P08015

Radiation effects on Single-Photon Avalanche Diodes manufactured in deep submicron CMOS technology

M Campajola[1,2*], F Di Capua[1,2], D Fiore[2,3], C Nappi[2,4], E Sarnelli[2,4], L Gasperini[5]
[1] Dipartimento di Fisica "E. Pancini", Università di Napoli "Federico II", Napoli, Italy
[2] Istituto Nazionale di Fisica Nucleare, Sezione di Napoli, Napoli, Italy
[3] Dipartimento di Ingegneria DIMES, Università della Calabria, Cosenza, Italy
[4] CNR-SPIN, Napoli, Italy
[5] Fondazione Bruno Kessler, Trento, Italy

E-mail: macampajola@na.infn.it

Abstract. CMOS Single-Photon Avalanche Diodes (SPADs) have been introduced recently in many scientific applications. This paper reports on the performance, in terms of Dark Count Rate (DCR), of a photo-sensor based on CMOS SPADs. The device has been subjected to an accurate investigation, in order to evaluate its behaviour in a radiation environment. Several irradiation tests were conducted, and a complete survey of their effects on the DCR behaviour has been performed. An overall increase in the DCR level has been measured, meaning that new defects have been introduced in the space charge region of the SPADs. Furthermore, for a fraction of the SPADs, DCR measurements show a Random Telegraph Signal (RTS) temporal pattern.

1. Introduction

Complementary metal-oxide-semiconductors (CMOS) Single-Photon Avalanche Diode (SPAD) have been introduced recently in many applications, ranging from industrial to the scientific ones. [1]. CMOS SPADs offer the high performance needed in application requiring the detection of weak light signal at the photon-counting level [2].

Although SPADs have been widely produced in custom process with excellent results, the CMOS technology offers the possibility to improve the detector response. Indeed, it allows the implementation of SPADs, front-end and signal processing electronics on the same substrate, improving the spatial and the temporal performances of the devices, and allowing signal processing inside the detector itself. Furthermore, CMOS technology allows the low-cost mass production needed to obtain large area sensors suitable for several research applications.

Currently, many experiments in high energy physics, medical physics and space applications took the challenge of developing dense circuits integrated into the detector itself, taking advantage of the maturity of CMOS technology. Beyond the engineering challenge of implementing complex circuits inside the detector, the real challenge is to create competitive devices with high performances and low noise [3]. Compared to custom devices, CMOS SPADs suffer from the high level of dark signals, expressed in term of Dark Count Rate (DCR). This depends on higher concentration of defects and

* Corresponding author

contaminants in the adopted substrate and a less accuracy in the CMOS fabrication process compared to the custom one. Defects concentration can be altered also after the fabrication steps due to environment factor. This is the case of devices operating in a radiation environment.
In this paper we report the results of the dark signal characterization on a SPAD sensor, developed in a deep submicron CMOS process. Several irradiation tests were conducted in order to test the radiation hardness of the devices. The dark count rate as well as the breakdown voltages (V_{br}) of the detector have been measured before and after proton irradiation. An overall increase in the DCR level has been measured, suggesting that new defects have been introduced in the space charge region of the SPADs. Furthermore, for a fraction of the SPADs, DCR measurements show a Random Telegraph Signal (RTS) temporal pattern.

2. Materials and methods

2.1. CMOS sensor developed by FBK

The Device Under Test (DUT) is a fully CMOS sensor designed by Fondazione Bruno Kessler (FBK) and implemented in a 150-nm CMOS technology. It is part of a test chips production, purposely made for the study of CMOS SPAD performances [4]. It implements several hundred of SPADs of different dimension according to Table 1

Table 1. SPADs design characteristics

Optical Window (μm)	5	10	15	20
Optical Area (μm^2)	22	97	222	397
Pitch Size (μm)	10.6	15.6	20.6	25.6

The chip contains SPADs organized in matrix groups of 5x5 elements and in linear arrays of 20 elements. Each SPADs is integrated together with its own front-end electronic. The basic design of the pixel architecture is reported on figure 1. SPADs are connected to a quenching transistor, an enable transistor and a Schmitt comparator, that digitalizes the pulse from the photodiode. Each pixel shares the same output and can be individually addressed through a decoder system.

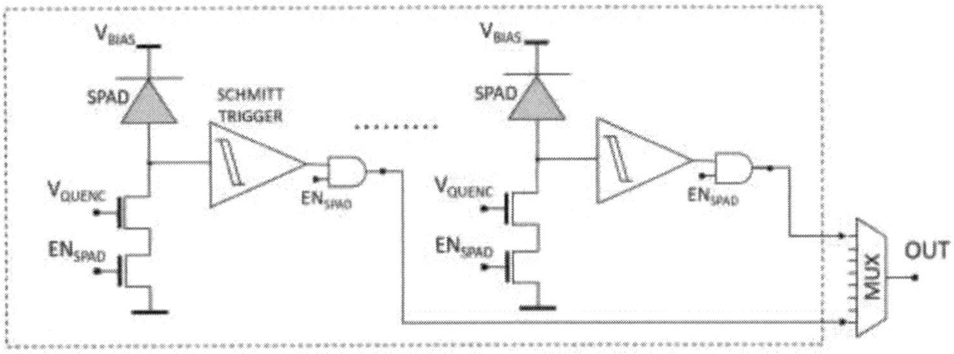

Figure 1. Pixel read-out circuit implemented inside the CMOS SPADs sensor.

2.2. Experimental setup

IOP Conf. Series: Journal of Physics: Conf. Series **1226** (2019) 012007 doi:10.1088/1742-6596/1226/1/012007

The purpose of this work was the study of the dark signal performances of the detector. To this purpose a fully automatized setup has been developed.
Measurements have been performed in a dark environment at a controlled temperature by means of a climatic chamber. The DCR measurement was conducted by means of an external counter connected to the chip output (figure 2). A dedicated software drives the setup instruments [5], so that automatic measurement of DCR have been performed as a function of the bias voltage V_{bias}.

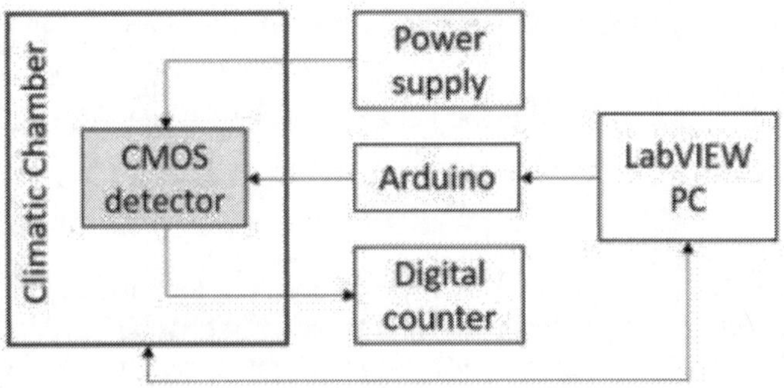

Figure 2. Schematic diagram of the experimental setup realized for the DCR measurements.

2.3. The proton facility
The proton beam tests have been performed at the Laboratori Nazionali del Sud (LNS) in Catania, by means of the tandem accelerator.
The irradiation tests have been performed with a proton beam of 24 MeV extracted in air. The actual energy of the beam on the DUT has been simulated through the FLUKA Montecarlo software, and found to be 20.5 MeV [5].
The beam dosimetry has been performed by means of a calibrated ionization chamber placed along the beam line. The beam profile, measured by means of a radio-chromic film placed in front of the device, showed a non-uniformity on the chip surface well below 10%. An integrated fluence of 5.63×10^{10} protons/cm^2 has been delivered, corresponding to a Displacement Damage Dose (DDD) of 376 TeV/g. According to [6], such radiation level is greater than that expected for typical space mission applications, but any case below the typical levels in high energy physics collider.

3. Results and discussion

3.1. Dark signal characterization
The device has been subjected to an accurate investigation of the dark count rate behaviour. Figure 3 shows the $DCR - V_{bias}$ plot for a subsample of 20 SPADs before irradiation test. At an overvoltage of 3.3 V, the mean DCR is of the order of the kHz. Breakdown voltage was extracted by subtracting the comparator reference voltage V_{ref} from the transition voltage in the DCR-V_{bias} plots. Figure 4 shows that the mean breakdown voltage value is about 23.5 V.
After irradiation, the DCR measurement have been performed in order to evaluate the irradiation effects on the CMOS SPADs sensor. We characterize DCR on several hundred of SPADs belonging to 2 different chips. While breakdown measures do not show significant changes after irradiation, the situation is more serious for the pedestal of DCR.

Figure 5 shows DCR measurements performed before irradiation, after each run and, finally, a month after the last irradiation. As shown the damage induced by radiation is not spatially uniform: some pixels show a DCR increase of more than two order of magnitude.

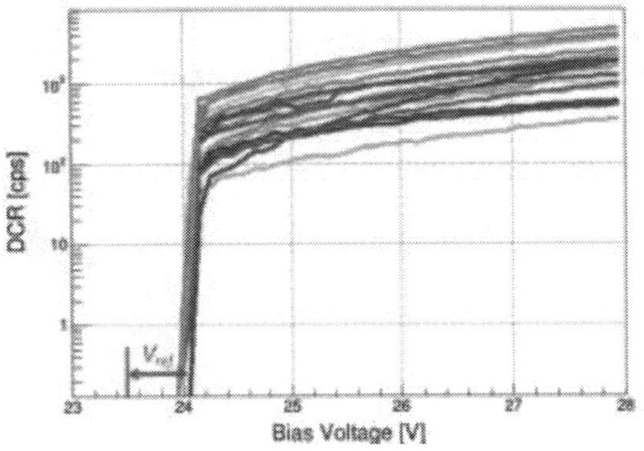

Figure 3. *DCR vs V_{bias}* on a sample of 20 CMOS SPADs .

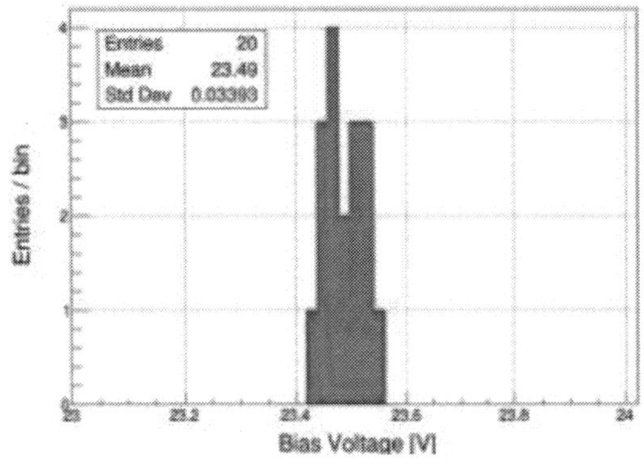

Figure 4. Breakdown voltage distribution measured on a sample of 20 SPADs.

Measurements performed a month after the irradiation test, show that a partial recover of the DCR has occurred, being the chip at room temperature.

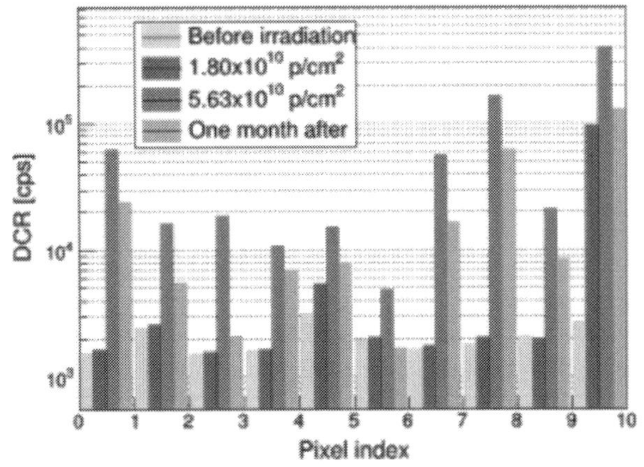

Figure 5. DCR variation after irradiation on a sample of 10 SPADs.

3.2. Random Telegraph Signal noise

After irradiation, many long-time DCR acquisitions on single SPADs showed in many pixels, discrete temporal fluctuations of the DCR between different levels.

The percentage of pixels affected by this behaviour after irradiation, depending on the SPADs layout, has found to be even more than 50%. Before irradiation, only a small percentage of SPADs is interested by this behaviour, while for almost all SPADs a stable level of DCR is observed.

To clearly present the phenomenon, a DCR measurement frame of few hours on a single SPAD and the corresponding DCR histogram are presented in figure 6 and figure 7. The DCR is observed to randomly switch between different DCR levels and four normal distributions of DCR can be observed.

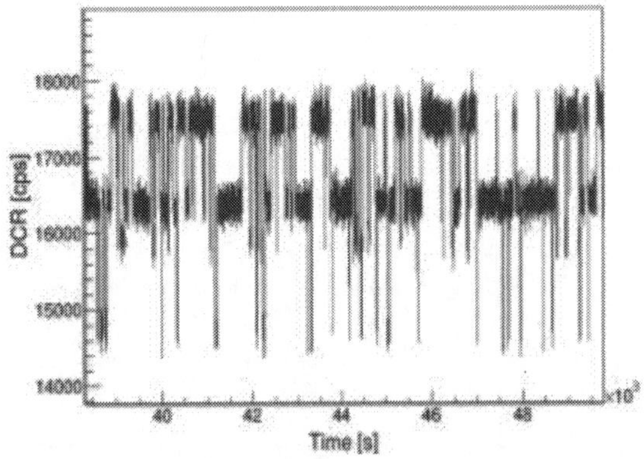

Figure 6. *DCR vs Time* in an irradiated SPAD affected by RTS.

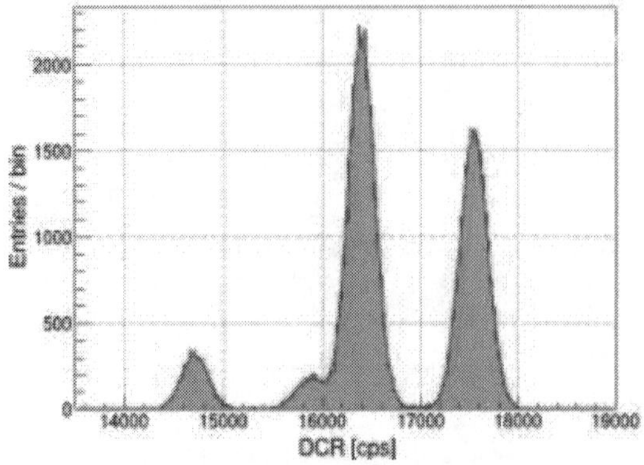

Figure 7. Distribution of the dark count rate. Four distinct distribution peak can be observed.

3.3. Discussion

In silicon devices, DCR is closely linked to the presence of defects and impurities in depletion region according to many mechanisms [7][8][9].

Due to a mechanism known as displacement damage, the concentration of such defects can increase during the detector lifetime if operated in a radiation environment.

Displacement damage can affect the detectors performances in many ways, depending on induced defect concentration, energy level and the respective electron and hole capture cross-section [10].

The observed results, can be related to the introduction of new defects with energy levels near the middle of the gap of the devices. According to [9], induced defects, acting as recombination/generation centres facilitate the transition of electrons and holes between bands and thus are responsible for an increase of the reverse current.

Irradiated SPADs showed large fluctuations in the DCR values, well beyond the non-uniformity of the proton beam. This behaviour can be addressed to the relatively low proton fluences and the small SPADs sensitive volume, that results in a large defects formation variability, dominated by statistic effects.

Beside the DCR pedestal increase, long time measurements, showed after irradiation a significant fraction of pixels affected by RTS pattern. Given that the increase in the DCR itself in a photon sensor is dangerous for its functionality, the RTS phenomenon can have a large impact on the noise performance of the sensor.

Post-irradiation measurements showed a large fraction of pixels affected by RTS behaviour. Several mechanisms have been proposed in literature in order to explain RTS in silicon image sensors [11][12][13]. The observed random switching of the DCR between two or more well-defined rates, suggests the presence of a metastable-stable defects located in the space charge region of the device.

4. Conclusion

The purpose of this paper was the understanding of the effects induced by radiation on a CMOS SPAD sensor. At this purpose, an irradiation campaign test has been performed with protons of 20.5 MeV, reaching a DDD of 376 TeV/g.

DCR measurements have been performed before and after the irradiation campaign. The samples demonstrated a significant increase in the DCR pedestal for most of the irradiated pixels, meaning that SPADs are quite sensitive to radiation damage.

Furthermore, several peaks were observed in the DCR distributions of single irradiated SPADs, suggesting the presence of radiation-induced metastable bulk defects.

IOP Conf. Series: Journal of Physics: Conf. Series **1226** (2019) 012007 doi:10.1088/1742-6596/1226/1/012007

References

[1] Palubiak D P and Deen M J 2014 CMOS SPADs: Design Issues and Research Challenges for Detectors, Circuits, and Arrays *IEEE J Sel Topics Quantum Electron* **20**

[2] Cova S, Ghioni M, Lacaita A, et al. 1996 Avalanche photodiodes and quenching circuits for single-photon detection *Appl. Opt.* **35** 1956

[3] Rochas A, Pauchard A R, Besse P, Pantic D, Prijic Z and Popovic R S 2002 Low-noise silicon avalanche photodiodes fabricated in conventional CMOS technologies *IEEE Trans. on Electron Devices* **49** 387-394

[4] Pancheri L and Stoppa D 2011 Low-noise Single-Photon Avalanche Diode in 0.15 μm CMOS Technology *Proc. of the European Solid-State Device Research Conf.* 179-182

[5] Campajola M 2017 Noise characterization of Single-Photon Avalanche Diodes M.A. Thesis (Napoli: University of Naples)

[6] Samwell S W, Hady A A, Mikhaill J S, Ibrahim M and Hanna Y S 2008 Studying the Total Ionizing Dose and Displacement Damage Dose effects for various orbital trajectories *Proceedings of IAU MEARIM*

[7] Shockley W and Read R T 1952 Statistics of the recombination of holes and electrons *Phys. Rev.* **87** 835–842

[8] Vincent G, Chantre A and Bois D 1979 Electric Field Effect on the Thermal Emission of Traps in Semiconductor Junctions J. *Appl. Phys.* **50** 5484-5487

[9] Xu Y, Xiang P, Xie X and Huang Y 2016 A new modelling and simulation method for important statistical performance prediction of single photon avalanche diode detectors *Semiconductor Science and Technology* **31**

[10] Srour J R and Palko J W 2013 Displacement Damage Effects in Irradiated Semiconductor Devices *IEEE Trans. Nucl. Sci.* **60** 1740–66

[11] Hopkinson G R, Goiffon V and Mohammadzadeh A 2008 Random telegraph signals in proton irradiated CCDs and APS *IEEE Trans. Nucl. Sci.* **55** 2197–2204

[12] Bogaerts J, Dierickx B and Mertens R 2002 Random telegraph signals in a radiation-hardened CMOS active pixel sensor *IEEE Trans. Nucl. Sci.* **49** 249–257

[13] Goiffon V, Magnan P et al. 2011 Evidence of a novel source of random telegraph signal in CMOS image sensors *IEEE Electron Device Lett.* **32** 773–775

IOP Conf. Series: Journal of Physics: Conf. Series **1226** (2019) 012008 doi:10.1088/1742-6596/1226/1/012008

Microphotoluminescence (μPL) measurements of bidimensional materials in a custom-made setup

F.V. Di Girolamo[1,2,*], *A. Di Lieto*[1,3], *A. Sottile*[3], *S. Roddaro*[1,3], *M. Tonelli*[1,3], *A. Tredicucci*[1,3]

1 Dipartimento di Fisica, Università di Pisa, Largo B. Pontecorvo 3, 56127 Pisa, Italy,
2 Istituto Nazionale di Fisica Nucleare, Largo B. Pontecorvo 3, 56127 Pisa, Italy.
3 NEST, Istituto Nanoscienze CNR, Piazza S. Silvestro 12, Pisa, 56127, Italy.

E-mail: `*flavia.digirolamo@df.unipi.it`

Abstract. The continuously growing interest in the so called bidimensional (2D) materials has been accompanied by a parallel development of techniques specifically devoted to manipulating and measuring micro-sized objects. Here a flexible microphotoluminescence (μPL) custom-made setup has been built on purpose: every setup component has been optimized for 2D materials; an automatized system able to perform measurements in different positions of the sample and execute mapping have been made; the dependence of the μPL intensity, peak center and other properties such as the Full Width Half Maximum (FWHM) on the position has been then extracted. The attention has been focused on WS_2 and phosphorene. Optical response has been compared between Chemical Vapor Deposition (CVD) grown WS_2 transferred on graphene and sapphire, especially focusing on the changes in the trion peak. Signatures have been found that can be attributed to photoluminescence from thick (more than five layers) phosphorene flakes.

1. Introduction

The discovery of graphene has paved the way for the utilization of the so called bidimensional (2D) materials: graphene exhibits exceptional transport, thermal, mechanical properties, but the lack of a gap restricts its field of application [1]; among others, transition metal dichalcogenides and phosphorene are instead largely used both for electronic and optical applications [1][2][3][4]. They can be obtained by exfoliation but also by Chemical Vapor Deposition (CVD), eventually accompanied by transfer on a particular substrate [5].
Transition metal dichalcogenides have a gap in the visible range: WS_2, in particular, has a gap of 2.1 eV (for single layer[1]) which becomes indirect and shrinks with increasing thickness [1][6]. Phosphorene represents instead the ideal step in between graphene and metal dichalcogenides, exhibiting a gap (1 eV for free standing single layer[3]) going towards the near infrared [7]. Phosphorene gap is always direct [7] and increases with decreasing thickness. It is characterized by a strongly anisotropic structure, which justifies the interest from both a teoretical and an applicative point of view [4].
A flexible custom-made setup has been built on purpose with the aim to perform microphotoluminescence (μPL) measurements on materials exhibiting optical properties lying in a very large spectrum region: every setup component has been optimized, and, in particular, the appropriate monochromator and detector were chosen for each material studied.

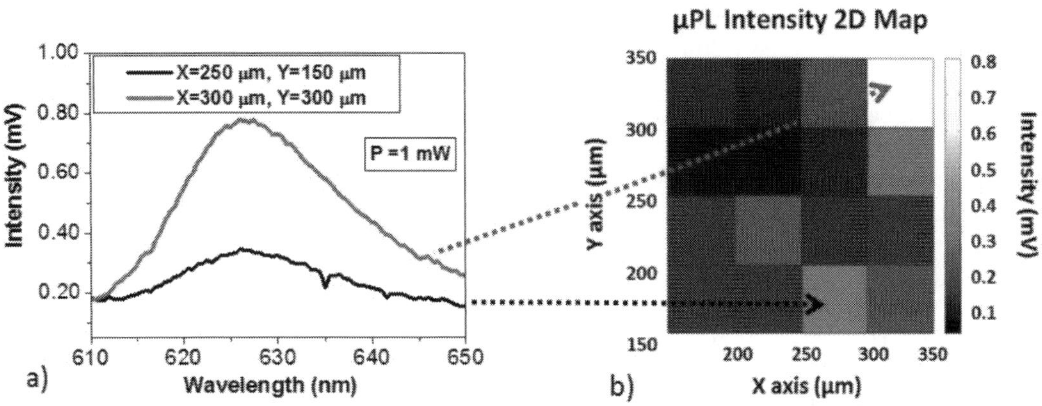

Figure 1. (a) Spectra of WS_2 μPL acquired in two different positions of the sample and (b) the corresponding intensity map at 1 mW. The arrows highlight the correspondence between the measurements and the points on the map.

Microtranslators allowed to focus and to move on the sample with precision (i.e. few microns) even in presence of small flakes. Moreover, they gave the chance to build up a specific system capable of performing automatized μPL measurements in different positions of the sample (e.g. see figure 1 (a)) and executing mapping. Maps have been obtained by fitting with one or more Lorentzian curves the μPL peaks relative to different positions and then extracting the intensity (e.g. see figure 1 (b)), the peak center and other properties such as the Full Width Half Maximum (FWHM). Microphotoluminescence measurements and maps have been performed on different bidimensional materials, such as CVD grown WS_2 on sapphire and graphene and exfoliated phosphorene on SiO_2. The optical properties have been studied in several different conditions, focusing on the peculiar features of the considered materials.

2. Materials and methods

WS_2, phosphorene and BaY_2F_8 : $10\%Er^{3+}$ doped samples have been gently provided respectively by Dr. C. Coletti, Dr. M.S. Vitiello, Prof. Tonelli and their research groups. More details on the experimental materials and methods can be found in [4][5][8].

The main elements of the flexible custom made μPL setup are the following: green solid state laser (DJ532-40 Thorlabs) equipped with a cooled laser mount (TCDLM9 Thorlabs); 100x microscope (M-Plan Apo NIR 100x); lock-in (SR830); three microtranslation stages (two M-111 and a M-126, Mercury) assembled in a cascade system and controlled by three motor controllers (C863 and C862, Mercury); camera (Basler acA640-100gm); clean-up filter (Techspec OD-6 laser line), CWL 532 nm and Raman longpass edge filter (Techspec). The detection system has been changed from time to time and it will be clearly identified.

Since the samples were sensitive to environmental conditions (especially phosphorene [3]), a cleaning procedure (i.e. with toluene and nitrogen) has been developed, along with a system to perform and maintain vacuum (around 100 mbar) during measurements.

Before performing photoluminescence measurements on bidimensional materials, the setup has been tested on a well-known bulk material (BaY_2F_8 : $10\%Er^{3+}$ doped). After that, microphotoluminescence (μPL) maps have been performed on different CVD grown WS_2 on sapphire samples in order to test the setup. CVD grown samples have been preferred for the large measurable area available. Sapphire has been preferred as a substrate compared to graphene because of μPL intensity quenching [9] (as will be discussed more widely in the

IOP Conf. Series: Journal of Physics: Conf. Series **1226** (2019) 012008 doi:10.1088/1742-6596/1226/1/012008

following section). In figure 1, μPL map results have been shown: in figure 1 (a) two μPL measurements collected in two different sample positions are reported while in figure 1 (b) the corresponding μPL intensity map is shown. The monochromator was the HR250 Jobin Yvon; the photomultiplier Hamamatsu R943-02; cobalt attenuation filters were in place.

3. Discussion

3.1. Optical properties of WS_2 on graphene and sapphire

The interest in graphene is not only limited to the material in itself, but also as a substrate for other bidimensional materials, composing heterostructures with peculiar properties [2][16]. Tan et al. [9] have instead investigated the effect of the high carrier conductivity of graphene used as substrate for WS_2: due to the migration of electrons in the underlying graphene, a quench in photoluminescence intensity has been experienced, accompanied by a decrease in the ratio of exciton/trion emission. The negative trion is a quasiparticle resulting from the combination of two electrons and a hole [10][11]. It came out to be an almost ubiquitous feature in bidimensional materials and, in particular, in transition metal dichalcogenides [2].

In our work, the optical properties of CVD grown single layer WS_2 on sapphire and on graphene have been compared, especially focusing on the trion peak. Mesurements have been performed on the selected samples using QEpro or a HR250 and a Hamamatsu R943-02. In figure 2, μPL curves measured have been shown. Both the samples have been measured immediately after fabrication and after 9 months. Best fits with two Lorentzian curves have been performed: the feature at lower wavelengths has been attributed to the neutral exciton (X), while the one at higher wavelengths to the negative trion (X^-) [9][12][13][14]. WS_2 on sapphire exhibits an exciton peak located at 620 nm, while on graphene it is located at 612 nm. Those values are generally in agreement with the literature, taking into account that variation between 620 and 640 nm can be found even in the same sample [6] and that a similar shift has been already measured on graphene [9]. Shifts in the position and in the Full Width Half Maximum (FWHM) can be attributed to several different factors, such as strain, doping, superficial defects [15][16]. The ratio of exciton/trion emission is higher for the sample on graphene, as expected [9]. The energy difference between the neutral exciton and the negative trion is in both cases around 30 meV (in agreement with literature [2][17]), irrespective of the different substrate.

The two samples have been measured several months after fabrication, and the spectra have been fitted with two Lorentzian curves. The trion feature is ubiquitously present in the sample on sapphire, even after 12 months from fabrication; on the other hand, in the sample on graphene the contribution of the trion after 9 months results negligible. Several factors can aid or hinder trion formation, such as thickness [18] or native doping [14]; nevertheless, it is probable that native doping during fabrication was comparable for the two samples. It is therefore reasonable taking into account the effects of the high conductivity of graphene in combination with WS_2, and concluding that electron migration in the underlying substrate could be the preferential mechanism for trion annihilation in WS_2 on graphene.

3.2. Microphotoluminescence measurements on a thick phosphorene sample

Photoluminescence measurements on phosphorene are available in the literature for flakes thickness comprises between 1 and 5 layers [3][7][19][20] . Exceeding this limit means that gap goes into the infrared range and that intensity drops. For 6 or more layers phosphorene flakes, IR absorption was then used for investigating optical and thus electronic properties, even under the application of stress [7].

Here several thick (i.e. thicker than 5 layers) flakes have been measured. A HR250 and a nitrogen cooled InSb detector have been used. Since the μPL intensity was expected to be low [19] measurements at high laser power have been performed, applying two different power intensites. Since measurements were performed under vacuum, damaging due to oxidation can

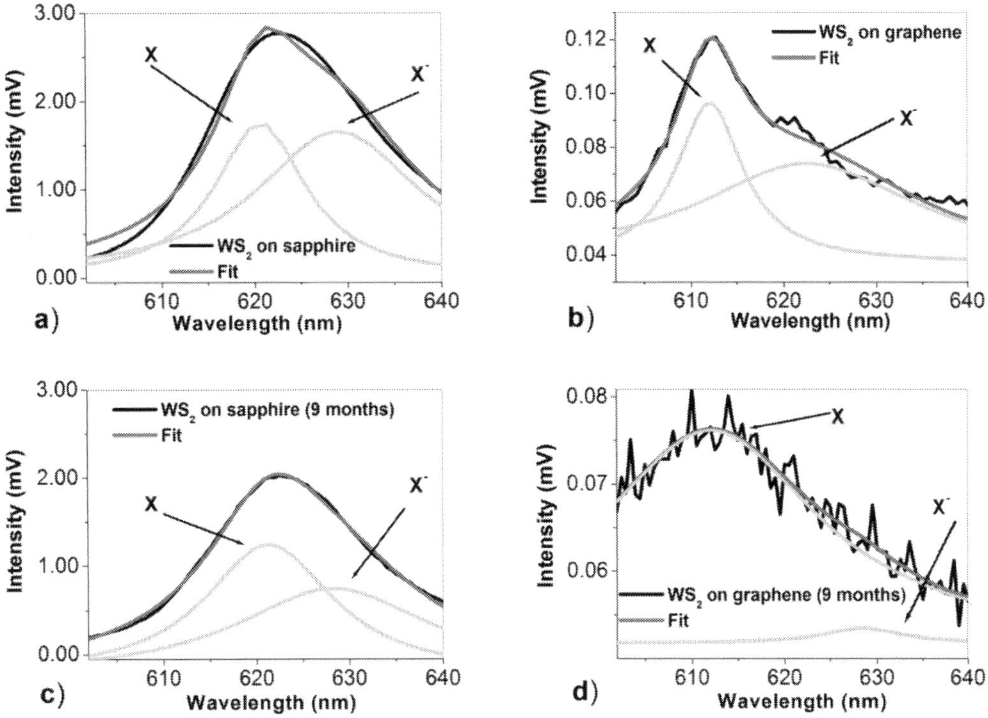

Figure 2. Microphotoluminescence curves measured on: WS_2 (a) on sapphire and (b) on graphene immediately after fabrication and (c) on sapphire and (d) on graphene after 9 months. Best fit with two Lorentzian curves evidence the neutral exciton (X) and the negative trion (X^-) peak.

be excluded [20].

In figure 3 hints are illustrated with the presence of a structure which tends to maintain its characteristics irrespective of the random nature of noise. In figure 3(a), the original spectra at different laser power and, in the inset, the flake picture. A 10 point smoothing was performed with three different smoothing techniques (FFT filter, adjacent averaging, Savitzky-Golay) and then fitted with two Lorentzian curves, with good agreement between them. In figure 3(b), the smoothed curves (with FFT filter) and the related Lorentzian fits are shown. It is possible to notice that the peak maintains its position irrespective of the increasing power, while the intensity increases, as expected [21]. Two features can be evidenced, respectively labeled as A (lower wavelength, lower intensity) and B (higher wavelength, higher intensity).

In bidimensional materials features at higher and lower energies than the gap have been widely reported [21] [22] and generally attributed to not intentional doping [7], to disorder or structural defects [2], to transitions between the lowest unoccupied molecular orbital (LUMO) and occupied molecular orbitals below the highest occupied molecular orbital (e.g. H-1 and H-2) [22]. The structure undergoes a slight shift at increasing power, which is in agreement with literature data on single layered and probably due to local laser heating [21]. Analogously, the not linear increasing of the μPL intensity with power is also expected when high power is applied [21], as in our case.

μPL results have been compared with the data available in the literature [19] (figure 3(c)), evidencing that the wavelength value of 1780 nm can be attributed to a 7 layers flake. In order

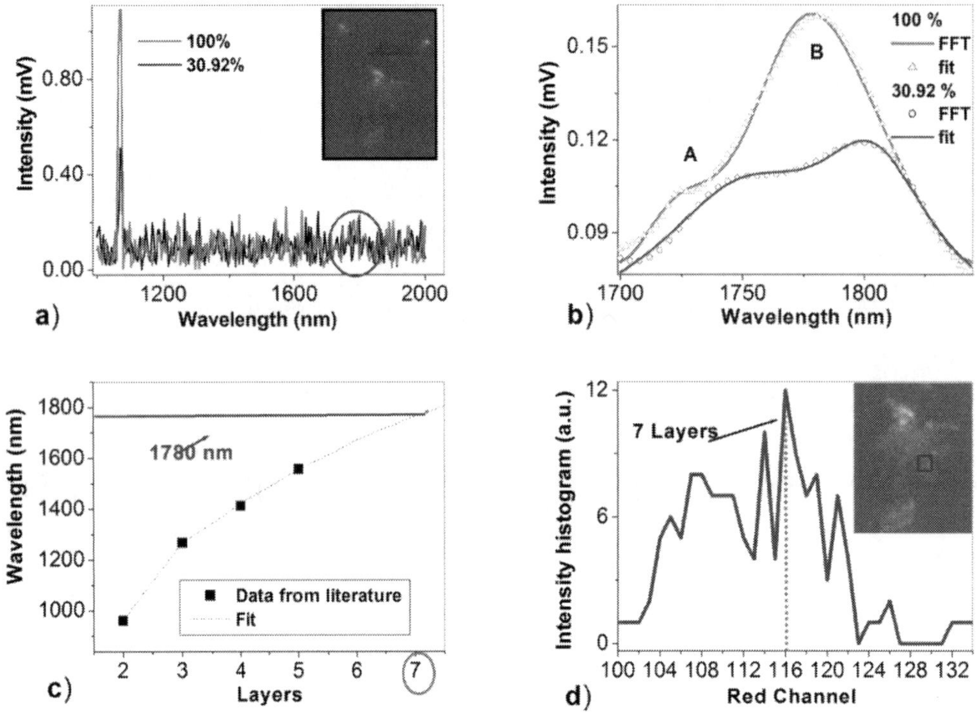

Figure 3. (a) μPL spectra performed under vacuum at 30.92% and 100% of laser power of the flake showed in the inset. The picture of the flake was taken in-situ in the measurement setup, under vacuum. (b) 20 pt smoothing with FFT filter and fits. (c) Comparison with the literature data [19]. (d) Histogram of red channel intensity of the measured flake region.

to get more information regarding the thickness of the measured flake and verify the previous attribution, the optical image of the flake taken in situ has been considered. In the literature, optical contrast of flakes picture is a parameter routinely used for thickness determination [23][24]. Castellanos Gomez et al. [20] reported a similar procedure in which histograms of several different regions of the flake have been performed. Here, histograms of different regions of the flake have been extracted from the red channel picture. Histograms have been taken of the sole substrate, of a region almost completely single layer and of other regions in which multiple layers are present. From them, a correspondence between the intensity peak red channel value and the thickness has been obtained. As expected, the red value of increasing thicknesses is multiple at a constant value and no correspondence has been found between the substrate value and the flake ones. The main peak is located at red channel 116, the value which corresponds to 7 layers (figure 3(d)), in agreement with the comparison with the μPL data found in the literature.

4. Conclusion

μPL measurements have been performed on WS_2 and on phosphorene. In the first case, the spectra of WS_2 on sapphire and on graphene immediately after fabrication and after several months have been fitted with two Lorentzian curves and compared. The results suggest that electron migration plays a crucial role in the annihilation of the trion for WS_2 on graphene. In the second case, μPL measurements have been performed on a thick phosphorene flake: a structure whose characteristics are in agreement with the literature and which is compatible with a 7 layers phosphorene flake has been found. The position of the peak remains constant at increasing power while the intensity increases. A slight shift and the not linear intensity increase are in agreement with the literature. Comparison with μPL data from the literature [19] suggests that the flake thickness could correspond to 7 layers, in agreement with the attribution of the thickness from histograms extracted from the optical image.

Acknowledgments
Authors wishing to thank F. Torri, A. Masetti, I.Grassini, G. Palumbo. Authors wishing to thank Dr. C. Coletti, Dr. M.S. Vitiello and Dr. L. Viti. We also acknowledge PRA (Università di Pisa) and ERC-IDEAS SOULMAN.

References

[1] Wang Q H, Kalantar-Zadeh K, Kis A, Coleman J N and Strano M S 2012 *Nat. Nanotech.* **7** 699.
[2] Pospischil A and Mueller T 2016 *Appl. Sci.* **6** 78.
[3] Han Liu, Neal A T, Zhen Zhu, Zhe Luo, Xianfan Xu, Tomanek D and Peide D. Ye 2014 *ACS Nano* **8** 4033.
[4] Viti L, Jin Hu, Coquillat D, Politano A, Knap W and Vitiello M S 2016 *Sc. Rep.* **6**: 20474.
[5] Rossi A, Bch H, Di Rienzo C, Miseikis V, Convertino D, Ameer Al-Temimy, Voliani V, Gemmi M, Piazza V and Coletti C, Scalable synthesis of WS_2 on graphene and h-BN: an all-2D platform for light-matter transduction.
[6] Gutierrez H R, Perea-Lopez N, Elas A L, Berkdemir A, Bei Wang, Ruitao Lv, F. Lopez- Uras, Crespi V H, Terrones H and Terrones M 2013 *Nano Lett.* **13**, 3447.
[7] Guowei Zhang, Shenyang Huang, Andrey Chaves, Chaoyu Song, V. Ongun zelik, Tony Low and Hugen Yan 2017 *Nat. Comm.* **8**:14071.
[8] Boccolini A, Faoro R, Favilla E, Veronesi S, and Tonelli 2013 *J. Appl. Phys.* **114** 064904.
[9] Haijie Tan, Ye Fan, Youmin Rong, Porter B, Chit Siong Lau, Yingqiu Zhou, Zhengyu He, Shanshan Wang, Harish Bhaskaran, and Warner J H 2016 *ACS Appl. Mater. Int.* **8**, 1644.
[10] Berkelbach T C, Hybertsen M S, and Reichman D R 2013 *Phys. Rev.* B **88**, 045318.
[11] Velizhanin K A and Saxena A 2015 *Phys. Rev.* B **92**, 195305.
[12] Zhengyu He, Wenshuo Xu, Yingqiu Zhou, Xiaochen Wang, Yuewen Sheng, Youmin Rong, Shaoqiang Guo, Junying Zhang, Smith J M, and Warner J H 2016 *ACS Nano* **10** 5847.
[13] Namphung Peimyoo, Weihuang Yang, Jingzhi Shang, Xiaonan Shen, Yanlong Wang, and Ting Yu 2014 *ACS Nano* **8**, 11320.
[14] Jingzhi Shang, Xiaonan Shen, Chunxiao Cong, Namphung Peimyoo, Bingchen Cao, Mustafa Eginligil, and Ting Yu 2015 *ACS nano* **9**, 647.
[15] Mlack J T, Paul Masih Das, Gopinath Danda, Yung-Chien Chou, Naylor C H, Zhong Lin, Lpez N P, Tianyi Zhang, Terrones M, Johnson A T C and Drndi M 2016 *Sci. Rep.* **7**:43037.
[16] Yuewen Sheng, Wenshuo Xu, Xiaochen Wang,a Zhengyu He, Youmin Ronga and Warner J H 2016 *Nanoscale* **8**, 2639.
[17] Ross J S, Sanfeng Wu, Hongyi Yu, Nirmal J. Ghimire, Jones A M, Grant Aivazian, Jiaqiang Yan, Mandrus D G, Di Xiao, Wang Yao and Xiaodong Xu 2013 *Nat. Comm.* **4**:1474.
[18] Weijie Zhao, Zohreh Ghorannevis, Leiqiang Chu, Minglin Toh, Kloc C, Ping-Heng Tan, and Goki Eda 2013 *ACS Nano* **7**,791.
[19] Shuang Zhang, Jiong Yang, Renjing Xu, Fan Wang, Weifeng Li, Muhammad Ghufran, Yong-Wei Zhang, Zongfu Yu, Gang Zhang, Qinghua Qin, and Yuerui Lu, 2014 *ACS Nano*, **8**, 9590 (2014).
[20] Castellanos-Gomez A, Vicarelli L, Prada E, Island J O, Narasimha-Acharya K L, Blanter S I, Groenendijk D J, Buscema M, Steele G A, Alvarez J V, Zandbergen H W, Palacios J J, and van der Zant H S J 2014 *2D Mater.* **1**, 025001.
[21] Surrente A, Mitioglu A A, Galkowski K, Tabis W, Maude D K, and Plochocka P 2016 *Phys. Rev.* B **93**, 121405(R).
[22] Shuaipeng Ge, Lisheng Zhang, Peijie Wang and Yan Fang 2016 *Sci. Rep.* **6**:27307.
[23] Likai Li, Jonghwan Kim, Chenhao Jin, Guojun Ye, Diana Y. Qiu, da Jornada F H, Zhiwen Shi, Long Chen, Zuocheng Zhang, Fangyuan Yang, Kenji Watanabe, Takashi Taniguchi, Wencai Ren, Steven G. Louie, Xianhui Chen, Yuanbo Zhang and Feng Wang, 2017 *Nat. Nanotech.*, **12**, 21.
[24] Meysam Akhtar, Anderson G, Rong Zhao, Adel Alruqi, Mroczkowska J E, Gamini Sumanasekera and Jasinski J B, 2017 *npj 2D Materials and Applications* **1**:5.

IOP Conf. Series: Journal of Physics: Conf. Series **1226** (2019) 012009 doi:10.1088/1742-6596/1226/1/012009

Multinucleation in the first-order phase transition of the 2d Potts model

F Corberi[1,2], L F Cugliandolo[3], M Esposito[2,3] and M Picco[3]

[1] INFN, Gruppo Collegato di Salerno, Unità di Salerno, Università di Salerno, via Giovanni Paolo II 132, 84084 Fisciano (SA), Italy.
[2] Dipartimento di Fisica "E. R. Caianiello", Università di Salerno, via Giovanni Paolo II 132, 84084 Fisciano (SA), Italy.
[3] Sorbonne Université, Laboratoire de Physique Théorique et Hautes Energies, CNRS UMR 7589, 4, Place Jussieu, Tour 13, 5ème étage, 75252 Paris Cedex 05, France.

E-mail: espoma@protonmail.ch

Abstract. Using large-scale numerical simulations we studied the kinetics of the 2d q-Potts model for $q > 4$ after a shallow subcritical quench from a high-temperature homogeneous configuration. This protocol drives the system across a first-order phase transition. The initial state is metastable after the quench and, for final temperatures close to the critical one, the system escapes from it via a multi-nucleation process. The ensuing relaxation towards equilibrium proceeds through coarsening with competition between the equivalent ground states. This process has been analyzed for different choices of the parameters such as the number of states and the final quench temperature.

1. Introduction

Phase transitions are a widespread phenomenon in physics, mathematics and in nature in general [1]. Some examples are melting ice, percolation, para- to ferro- transitions in magnetic systems, Bose-Einstein condensation, etc.

A system undergoes a phase transition when it changes its properties in a discontinuous way, that is, by exhibiting a discontinuity in a thermodynamic function. Specifically, we can make a distinction between two cases. When the first-order derivative of the Gibbs free energy $\mathcal{F}$ (i.e., the order parameter) is discontinuous, we say that a *first-order* phase transition is taking place. When the discontinuity affects second- or higher-order derivatives of $\mathcal{F}$, we name them *continuous* and we classify the former as *second-order*.

In order to better understand the basic features of these two kinds of phase transitions,

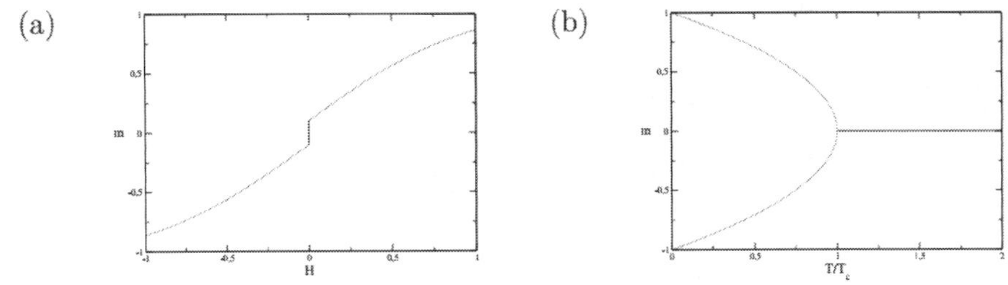

Figure 1: Magnetization density, $m = M/N$ of the Ising model, (a) as a function of magnetic field at $T < T_c$ (first-order phase transition), (b) as a function of temperature at zero magnetic field (second-order phase transition).

it is useful to consider the familiar Ising model, described by the Hamiltonian

$$H_I = -J \sum_{\langle ij \rangle} \sigma_i \sigma_j - H \sum_i \sigma_i , \tag{1}$$

where J is a coupling constant, the first sum is restricted to nearest-neighbors on a lattice, $\sigma_i = \pm 1$ is a Boolean spin variable, and H is an external magnetic field. The order parameter is the total magnetization

$$M = \sum_i \sigma_i , \tag{2}$$

which is the first derivative of $\mathcal{F}$ with respect to the magnetic field.

It is well-known that the 2d Ising model undergoes a first-order phase transition at a fixed $T < T_c$ when the sign of the magnetic field is switched. At null field, the model exhibits a second-order phase transition when it is cooled from $T_i > T_c$ to $T_f < T_c$.

The magnetization (normalized by the number of spins) is plotted in Figs. 1a and 1b for the two cases. It can be seen that m is discontinuous at $H = 0$ in the first-order phase transition and continuous in the second-order phase transition. In this last case, its first derivative – that is, the magnetic susceptibility – exhibits a discontinuity at $T = T_c$.

This is the overall picture when a control parameter is varied continuously and the system changes state via quasistatic transformations. If, on the other hand, the control parameter is varied abruptly, the system passes through non-equilibrium configurations until slow dynamical processes eventually lead it to equilibrium. Here we describe two of these processes that will be considered later, namely nucleation and coarsening. Coarsening characterizes the dynamics of second-order phase transitions [2, 3, 4], and it is the process whereby the 2d Ising model at $H = 0$ orders after a quench from $T_i > T_c$ to $T_f < T_c$. In this case, it consists of the growth of domains driven by surface tension, as depicted in Fig. 2. Specifically, bigger domains devour smaller ones so that the typical domain size $R(t)$ grows in time as $R(t) \sim t^{1/2}$. Since (at finite temperature) equilibrium is reached when $R(t) \sim L$, relaxation takes an infinite time in the thermodynamic limit.

Nucleation characterizes the dynamics of first-order phase transitions [5, 6, 7, 8, 9, 10]. It can take place in a system that is initially in a metastable state and has to exceed a

IOP Conf. Series: Journal of Physics: Conf. Series **1226** (2019) 012009 doi:10.1088/1742-6596/1226/1/012009

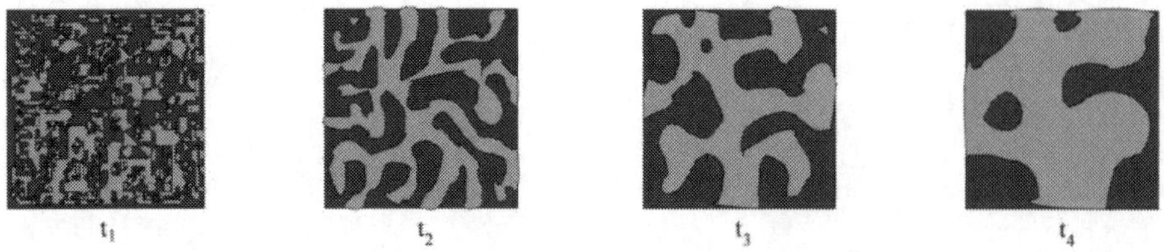

Figure 2: Pictorial representation of a system during coarsening at $t_1 < \cdots < t_4$.

free-energy barrier to reach equilibrium. The system tries to escape such metastable state via nuclei formation of the new phase into the old one. The contribution of a nucleus of size r to the free energy consist of a positive surface term and a negative bulk term $\Delta\mathcal{F} = -\Delta f\; r^d + s\; r^{d-1}$, where Δf is the bulk free energy, s is the surface tension and d is the dimensionality. Until a critical nucleus of size $r^* \propto s/\Delta f$ is created, the dynamics is inefficient as nuclei of a smaller size grow and shrink continuously without affecting the free energy. When the critical nucleus is created, the system has jumped over the free-energy barrier and small fluctuations start the relaxation towards the equilibrium state that can happen, for instance, via coarsening. The two processes, then, are not mutually exclusive, but can happen in the same system on different time scales.

The aim of this study is to characterize the dynamics following a sudden quench from initial conditions typical of equilibrium at $T_i = \infty$ to a temperature below the critical one such that the ordering process starts via a multinucleation of many competing phases and continues through coarsening. The simplest model that realises this phenomenology is the Potts model. We now proceed to define it and study it.

2. Results

The Potts model is described by the Hamiltonian [11, 12, 13]

$$H_P = -J \sum_{\langle ij \rangle} \delta(\sigma_i, \sigma_j) \;, \tag{3}$$

where $J > 0$ is a coupling constant, the sum is restricted to nearest-neighbors on a lattice, $\delta(a, b)$ is the Kronecker delta and σ can take integer values from 1 to $q \geq 2$. This model is a generalization of the Ising model with zero external magnetic field, to which it reduces for $q = 2$.

This model undergoes a phase transition at the critical temperature [12, 13]

$$T_c(q) = \frac{1}{\ln\left(1 + \sqrt{q}\right)} \;. \tag{4}$$

In particular, for $2 \leq q \leq 4$, the transition is of the second-order, while it is of the first-order for $q > 4$. The extent of the region of metastability can be determined with modern and powerful numerical methods [14, 15].

We defined the model on an $L \times L$ square lattice with periodic boundary conditions in the first-order phase transition regime, that is for $q > 4$. We studied its kinetics after a quench from $T_i > T_c$ to $T_f < T_c$, for different values of the final temperature T_f, the number of available states q and the system size L. Starting from a completely disordered configuration – corresponding to an infinite-temperature state – the dynamical evolution proceeds via a Monte Carlo Markov Chain [16], where a lattice spin and a number $l \in [1, q]$ are randomly picked and the spin is changed from state r to state l with the Metropolis transition rate

$$w_{rl} = \min\left(1, e^{-\beta \Delta E_{rl}}\right) , \tag{5}$$

where $\beta = T_f^{-1}$ (we set the Boltzmann constant to unity) and ΔE_{rl} is the energy difference between the configurations before and after the attempted spin flip. A Monte Carlo Step (MCS) is the time unit and corresponds to $N = L \times L$ spin flip attempts.

The dynamical evolution of the Potts model has been studied in previous works that were mainly focused on the coarsening regime [17, 18, 19, 20, 21, 22], (see discussion below). We directed instead our analysis to the nucleation process.

Our primary quantity of interest is the excess of energy density [17]

$$\phi_E(t) = e(t) - e(\infty) , \tag{6}$$

where $e(t)$ is the energy density at a time t after the quench and $e(\infty)$ is the equilibrium energy density. The latter can be computed starting from a completely ordered configuration, that is to say, all spins talking the same value among the q possible ones (equilibrium at zero temperature) and letting the system relax towards equilibrium at the desired temperature.

The behavior of the excess energy ϕ_E is plotted in Figs. 3a and 3b for $q = 5, 9$ and 100, for different sizes L and values of the final temperature T_f. First of all, it can be seen that for $T \lesssim T_c$, we can identify three different regimes. After a very short transient, ϕ_E reaches a long-lasting plateau. At a time $\tau_N(T_f)$ – which we call *nucleation time* – ϕ_E decreases abruptly to a value that depends on the final temperature T_f. This is consistent with the picture of nucleation given before, where the system is initially stuck in the metastable state and tries to escape via nuclei formation. The dynamics are inefficient until several critical nuclei are created around a time $\tau_N(T_f)$, which results in a constant value of ϕ_E. At $t = \tau_N$, the system exceeds the free-energy barrier. In Fig. 3a, the violet dashed curve represents the function $f(t) = 0.51 \exp(-5.98 \cdot 10^{-6}\, t)$, which is the best exponential fit for the fast decay at $T_f = 0.72$.

Relaxation well after $\tau_N(T_f)$ proceeds via coarsening, as can be seen by looking at the power-law behavior of the excess of energy $\phi_E \propto \lambda_q(T_f)\, t^{-1/2}$, which implies [2] $R(t) \propto t^{1/2}$, as expected in coarsening. Notice how the first-order phase transition is less evident for $q = 5$ (inset of Fig. 3a). This is probably due to the proximity to the case $q = 4$, where the transition turns second-order.

It is evident that both the shape of the fast decay of ϕ_E after τ_N and the temperature range in which nucleation dominates the dynamics depend on the value of q. Specifically, the rapid relaxation becomes sharper and is observed for a wider temperature range the greater the value of q. Let us notice that the value of the excess of energy in correspondence of the plateau ϕ_E^* depends on q, but it is rather temperature-independent. We have plotted in Fig. 5 the absolute value of ϕ_E^* as a function of the number of states.

IOP Conf. Series: Journal of Physics: Conf. Series **1226** (2019) 012009 doi:10.1088/1742-6596/1226/1/012009

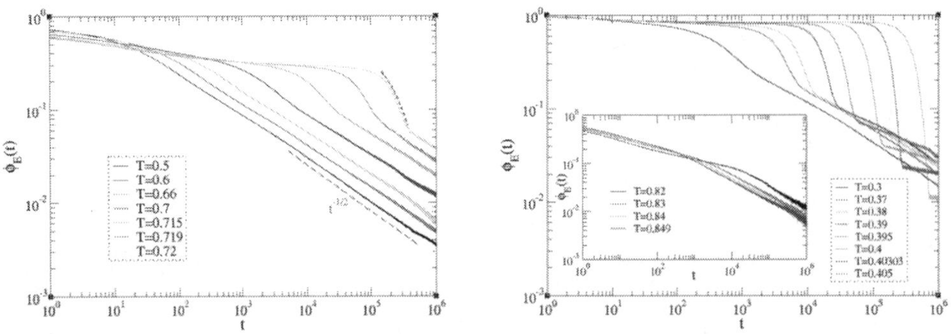

Figure 3: ϕ_E vs t for $q = 9$ ($T_c = 0.71235$), $L = 1000$ and different values of the temperature T_f that are given in the key (*Left Panel*). In the right panel, ϕ_E vs t for $q = 100$ ($T_c = 0.41703$), $L = 700$ and different values of T_f also in the key. In the inset, the same but for $q = 5$ ($T_c = 0.85153$).

For $q < 50$, the value of ϕ_E^* is well approximated by the logarithm of $q - 4$. For greater values of q one has saturation to a constant value $\phi_E^* \simeq 0.85$.

Four snapshots of the lattice for $q = 9$, $L = 1000$ and $T = 0.715$ are shown in Fig. 4 for the three different regimes of ϕ_E. At $t = 10^3$ – in the plateau region – the system is still in a completely disordered configuration, qualitatively similar to the one at $t = 0$, a sign that the dynamics is initially inefficient. Some larger domains are visible but they are too small to trigger nucleation. During the fast decay, nuclei of different phases grow very rapidly, see Fig. 4b. Coarsening, as shown in Figs. 4c and 4d is characterized by growth of bigger domains to the detriment of smaller ones. At $t = 10^6$, the time at which the snapshot in Fig. 4d was recorded, four phases are no longer present in the system.

3. Conclusions

In this work we investigated the dynamics of the Potts model for $q > 4$ after a quench from $T_i > T_c$ to $T_f < T_c$. Relaxation proceeds via multinucleation for $T_f \lesssim T_c$, followed by coarsening. This behavior has been characterized by looking at the excess of energy, that shows a temperature-independent plateau at the beginning, followed by a fast decay – which is best fitted with an exponential – and ends with a coarsening-like behavior $\phi_E \propto t^{-1/2}$. The behavior of the plateau energy $\phi_E^*(q)$ is consistent with the logarithm

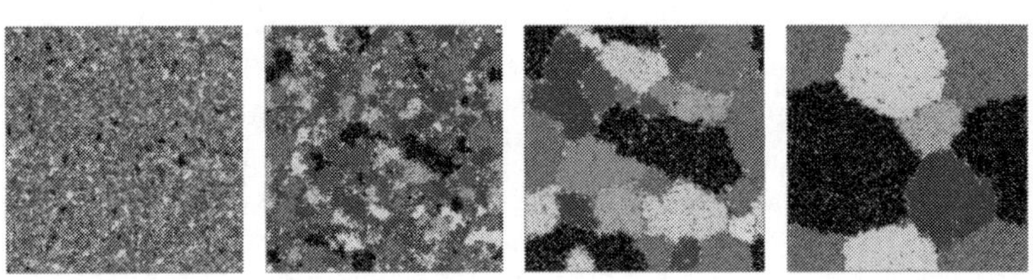

(a) (b) (c) (d)

Figure 4: Snapshots of the lattice at different times for $q = 9$, $L = 1000$ and $T = 0.715$ (orange curve in Fig. 3a) in the three stages of plateau (Fig. 4a), fast decay (Fig. 4b) and coarsening (Fig. 4c and Fig. 4d). Notice that – at $t = 10^6$ – four phases have been eliminated.

IOP Conf. Series: Journal of Physics: Conf. Series **1226** (2019) 012009 doi:10.1088/1742-6596/1226/1/012009

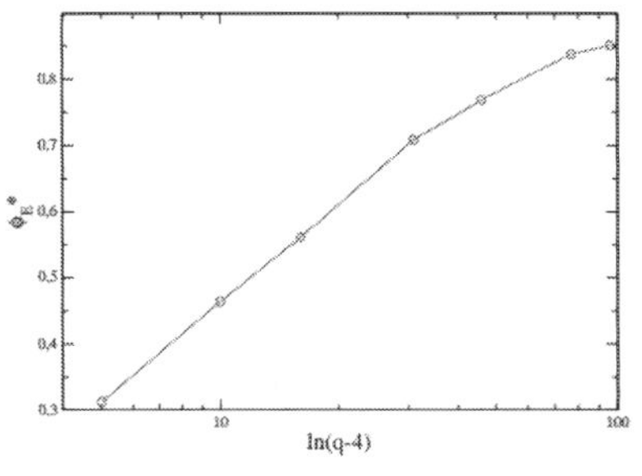

Figure 5: Value of ϕ_E on the plateau as a function of the logarithm of $q-4$.

of $q-4$. During coarsening, phases are eliminated. In the light of our results, it can be interesting to investigate the spontaneous symmetry breaking mechanism whereby this elimination takes place, and compare it with the coarsening process that characterizes the dynamics of the Ising model, to identify possible similarities and differences. It could also be useful to study the geometrical properties of the clusters to understand whether their evolution is consistent with the Neumann-Mullins law [23, 24].

Acknowledgements LFC wishes to thank F. Krzakala for early discussions on this problem. She is a member of Institut Universitaire de France.

References

[1] Papon P, Leblond J and Meijer P H E *The Physics of Phase Transitions, Springer-Verlag, 2006.*
[2] Bray A J Adv. Phys. 1994 **43** 357.
[3] Puri S *Kinetics of phase transitions*, ed. by Puri S and Wadhawan V (Taylor and Francis, 2009).
[4] Corberi F and Politi P Comptes Rendus de Physique 2015 **16** 255.
[5] Langer J S Ann. Phys. (N.Y.) 1967 **41** 108.
[6] Langer J S Phys. Rev. Lett. 1968 **21** 973.
[7] Langer J S Ann. Phys. (N.Y.) 1969 **54** 258.
[8] Binder K and Stauffer D Adv. Phys. 1976 **25** 343.
[9] K Binder K Rep. Prog. Phys. 1987 **50** 783.
[10] X C Zeng and D W Oxtoby J. Chem. Phys. **94**, 4472 (1991).
[11] Potts R *Doctoral thesis*, 1952.
[12] Wu F Y Rev. Mod. Phys. 1982 **54** 235.
[13] Baxter R J *Exactly solved models in statistical mechanics* 1982 Academic Press 1st edition.
[14] Loscar E S, Ferrero E E, Grigera T S and Cannas S A J. Chem. Phys. 2009 **131** 024120.
[15] Ferrero E E, De Francesco J P, Wolovick N and Cannas S A Comp. Phys. Comm. 2002 **183** 1578.
[16] Newman M E J and Barkema G T *Monte Carlo Methods in Statistical Physics* 1999 (Oxford: Clarendon Press)
[17] Ferrero E E and Cannas S A Phys. Rev. E 2007 **76** 031108.
[18] Ibáñez de Berganza M *et al* Eur. Phys. J. Sp. Topics 2007 **143** 273.
[19] Sanders D P, Larralde H and Leyvraz F Phys. Rev. B 2007 **75** 132101.
[20] Loureiro M P O, Arenzon J J, Cugliandolo L F, and Sicilia A Phys. Rev. E 2010 **81** 021129.
[21] Loureiro M P O, Arenzon J J, and Cugliandolo L F Phys. Rev. E 2012 **85** 021135.
[22] Olejarz J, Krapivsky P L, and Redner S J. Stat. Mech. 2013 P06018.
[23] von Neumann J Metal interfaces. Cleveland: American Society for Testing Materials 1952 108.
[24] Mullins W W J. Appl. Phys. 1956 **270** 900.

IOP Conf. Series: Journal of Physics: Conf. Series **1226** (2019) 012010 doi:10.1088/1742-6596/1226/1/012010

Influence of thermal conductivity and of non-constant relaxation time on thermoelectricity in Mg_3Sb_2

R Farris, M B Maccioni, A Filippetti, V Fiorentini

Department of Physics, University of Cagliari, Cittadella Universitaria, I-09042 Monserrato (CA), Italy

E-mail: roberta.farris@dsf.unica.it

Abstract. We discuss the effects of different approximations to thermal conductivity and relaxation time on the thermoelectric figure of merit ZT in n-doped Mg_3Sb_2. We use density-functional electronic structure and Bloch-Boltzmann transport theory with an energy-dependent relaxation time. ZT is monotonically increasing with T in our range (300-700 K) with a maximum value of ~1.5.

1. Introduction

Thermoelectrics make the recycling of thermal waste into usable energy a possibility. The material-specific thermoelectric figure of merit is

$$\mathrm{ZT} = \frac{\sigma \mathrm{S}^2 \mathrm{T}}{\kappa} = \frac{P\mathrm{T}}{\kappa_e + \kappa_\ell},$$

where σ is the electrical conductivity, κ, κ_e and κ_ℓ the total, electronic, and lattice thermal conductivities, S the Seebeck coefficient, P the power factor, and T the temperature. Mg_3Sb_2 and related compounds have emerged as interesting candidates [1] due to their small [2] lattice thermal conductivity and appreciable power factor.

2. Results

We use density functional theory in the generalized gradient approximation [3] and the projector augmented wave method implemented in the VASP code [4, 5] to obtain electronic bands on a (24, 24, 18) k-point grid in the Brillouin zone; we then plug the band structure into the calculation of Onsager coefficients in the linearized Bloch-Boltzmann approximation [6, 7] as implemented in the BoltzTrap [8] code in the rigid band and constant-relaxation-time approximation. We further employ analytical energy-dependent expressions for the relaxation time τ, including electron-acoustic phonon scattering, in the deformation potential approach; electron polar-optical scattering, in Ridley's formulation; charged impurity scattering, via the Brooks-Herring formula [9, 10]. These formulas, implemented in BoltzTrap, were previously used with satisfactory results [11, 12, 13]. The relevant parameters needed in the various scattering terms have been either calculated or imported from the literature. We assume n-type doping of 3.2×10^{19} cm^{-3} with monovalent impurities. This is the optimal doping for thermoelectricity

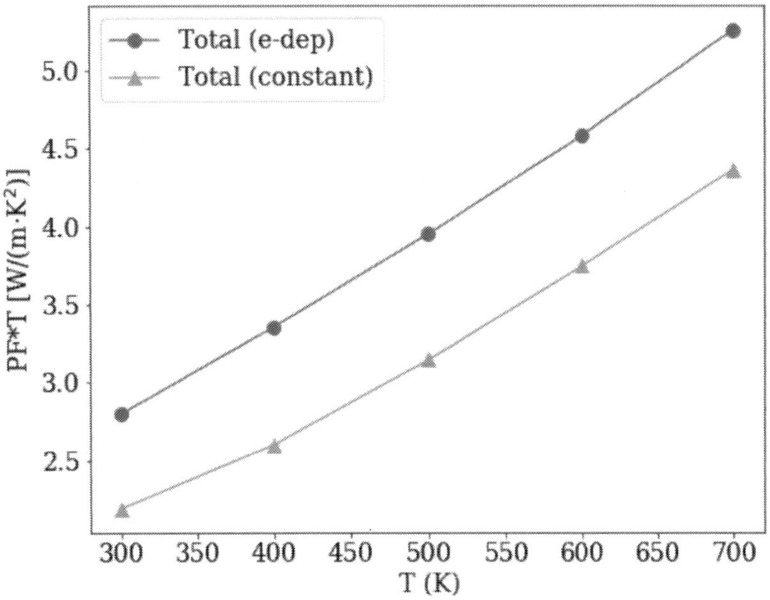

Figure 1. Power factor multiplied by T, computed with energy-dependent (●) and constant (△) scattering time.

as it maximizes ZT in this material [13]. We have previously calculated the lattice thermal conductivity [2], and we use the results below.

In Fig. 1 we show the power factor for energy-dependent and constant relaxation time. Clearly, due to the combination of effects on electrical conductivity and Seebeck coefficient, the power factor is significantly larger in the former case. The effect on ZT would be to lower it by about 20-25%. To determine ZT, we now consider the second key ingredient: thermal conductivity.

In Fig. 2, we display the total thermal conductivity obtained summing the electronic component calculated in the energy-dependent and constant time approximations to the lattice component [2], also displayed. It can be seen that a) the electronic component is as important as the lattice component, given the small size of the latter; and that b) the constant-time approximation overestimates significantly the thermal conductivity at high T, leading, in particular, to a sharp rise which is not observed in experiment [1]. Using this total κ, one eventually predicts a ZT about a factor 2 smaller. At this doping, the total κ still follows the downward trend in T of the lattice component.

The figure of merit ZT for our n-doped polycrystal, shown in Fig. 3 as filled blue squares, is essentially in agreement with experiment for the same kind of material (see Fig. 1a of Zhang *et al.*, Ref.[1]), with a maximum ZT=1.5 at the top of our T range. ZT in the constant-time approximation is about a factor of 2 smaller, due to the larger increase in κ_e with T in that case. Using the lattice thermal conductivity of the crystal, which is ([2], [14]) nearly tenfold the polycrystal's at room T, the figure of merit ZT is drastically suppressed.

3. Conclusions

In summary we examined the thermoelectric coefficients and figure of merit ZT in Mg_3Sb_2 with an energy-dependent scattering time. We find it is essential to include both the lattice and electronic components of thermal conductivity into the calculations, as they affect the final ZT significantly. In particular, we have demonstrated the effects of using the polycrystal vs the

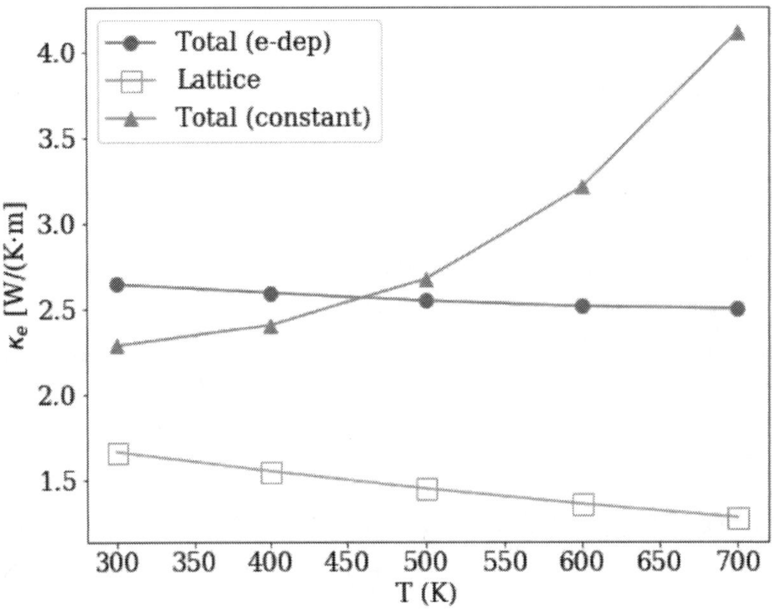

Figure 2. Total thermal conductivity computed with energy-dependent (●) and constant scattering time △for the polycrystal. The calculated lattice component [2] for the polycrystal is displayed for reference (□).

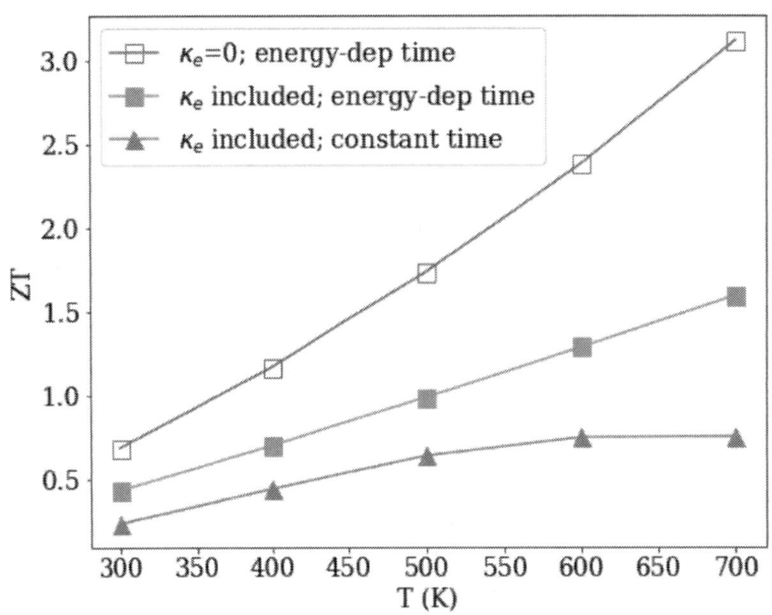

Figure 3. Various incarnations of the ZT figure of merit: full ZT with poly κ_ℓ and energy-dependent relaxation time (blue □), ZT with crystal κ_ℓ and energy-dependent relaxation time (orange ■) and ZT with constant scattering time and poly κ_ℓ (green △).

crystal κ_ℓ. Our final ZT is in good agreement with experiment.

4. Acknowledgments
Work supported in part by UniCA, Fondazione di Sardegna, and Regione Autonoma della Sardegna via Progetti biennali di ateneo 2016, by CINECA, Bologna, Italy through ISCRA grants, and CRS4 Computing Center, Pula, Italy.

References
[1] Zhang J, Song L, Hindborg Pedersen S, Yin H, Hung L T, and Brummerstedt I B, 2017 *Nature Comm.* **8** 13901
[2] Maccioni M B, Farris R, and Fiorentini V, 2018 *Phys. Rev.* B **98**, 220301
[3] Perdew J P, Burke K, and Ernzerhof M 1996 *Phys. Rev. Lett.* **77** 3865
[4] Kresse G and Furthmüller J 1996 *Phys. Rev.* B **54** 11169
[5] Kresse G and Joubert D. 1999 *Phys. Rev.* B **59** 1758
[6] Allen PB 1978 *Phys. Rev.* B **17** 3725
[7] Allen P B, Pickett W, and Krakauer H, 1998 *Phys. Rev.* B **37**, 7482
[8] Madsen G K H and Singh D J 2006 *Comput. Phys. Commun.* **175** 67
[9] Balkanski M and Wallis R F 2000 *Semiconductor Physics and Applications* (New York: Oxford University Press
[10] Ridley B K 2000 *Quantum Processes in Semiconductors* (Oxford: Clarendon Press)
[11] Filippetti A *et al.* 2012 *Phys. Rev.* B **86** 195301
[12] Delugas P *et al.* 2013 *Phys. Rev.* B **88** 045310
[13] Farris R, Maccioni M B, Filippetti A and Fiorentini V, 2019 *J. Phys.: Condens. Matter* **31** 065702
[14] Li J, Zheng S, Fang T, Yue L, Zhang S and Lu G, 2018 *Phys. Chem. Chem. Phys.* **20** 7686

IOP Conf. Series: Journal of Physics: Conf. Series **1226** (2019) 012011 doi:10.1088/1742-6596/1226/1/012011

Graphenic carbon as etching mask: patterning with laser lithography and KOH etching

A Furio[1,2], M Stelzer[1], M Jung[1], H C Neitzert[2], F Kreupl[1]

[1] Technical University of Munich (TUM), Arcistr. 21, Munich, Germany

[2] University of Salerno, Via Giovanni Paolo II 132, Fisciano, Italy

E-mail: furioalfonso@gmail.com

Abstract. The wet anisotropic etching process is generally used in the field of micromachining (MEMS), particularly for commercial products such as accelerometers. Hard masks like oxide or nitride play a key role in the transfer of patterns to the substrate during the lithography process. This work reports on the use of polycrystalline graphenic carbon as an etch mask for wet chemical processing and outlines a simple method to create patterned structures on (100) silicon wafers. Graphenic carbon (GC) was deposited on the silicon substrate by chemical vapor deposition (CVD) using C_2H_4 as precursor. The desired pattern was written in the spin-coated negative photoresist using UV laser lithography. Different geometrical shapes were printed on the substrate with dimensions ranging from 10 to 50 micrometers. In the next stage, the O_2 plasma etched away the carbon from the area not covered by the photoresist, acting as an additional mask for this and the subsequent processing steps. Finally, the sample was immersed in the KOH bath saturated with isopropanol and the etching rate was evaluated for each crystal plane. Compared to the use of a sacrificial oxide mask, this technique is simpler and produces more reliable results.

1. Introduction

Anisotropic wet etching is one of the most used techniques for fabricating 3D structures as MEMS on silicon substrates. Particularly, bulk etching is widely used to remove silicon from the substrates in order to produce sensors or actuators. Typical micromechanical devices, manufactured on (100) silicon wafer, are cantilevers or membranes that usually need supporting structures based on oxides or nitrides [1]. These insulating hard masks also play an important role during the lithography step for transferring patterns to the substrates. The etching process is usually carried out in an aqueous solution of KOH, but as reported by Barycka et al. [2], the KOH solution saturated with isopropyl alcohol is the most frequently used to obtain a high etch rate ratio and smooth surface on (100) silicon. The choice of the etching solution depends on different factors such as etch rate and anisotropy. The etch rate differs between the crystal planes and the concentration of the used solution [3, 4]. The aim of this work is to create 3D structures on (100) silicon wafer using photolithography and wet anisotropic etching in a commonly used solution of KOH and isopropanol. In this context, the use of graphenic carbon (GC) [5, 6] as an etching mask instead of sacrificial oxide is a new feature. As reported in [7, 8], it is possible to form GC membranes on one side of a wafer exploiting the undercut created by the KOH (saturated with isopropanol) on a silicon substrate. GC is a good candidate for the future process due to the fact that it is inert to most common solvents such as acetone and NMP

IOP Conf. Series: Journal of Physics: Conf. Series **1226** (2019) 012011 doi:10.1088/1742-6596/1226/1/012011

and, with a thickness of a few nanometers, it is resistant to HF and KOH. Another advantage is that it is extremely easy to remove with a O_2 or H_2 plasma and that it is compatible with the CMOS standard process through chemical vapor deposition (CVD). Moreover, the thickness of the GC is easily tunable by controlling the parameter of the deposition. The high-temperature stability, the low contact resistance to silicon and the endurance to high current densities make this material the ideal candidate for metal-silicon contacts [9]. In the process flow of this work, this means saving process steps using a conductive mask.

2. Sample preparation

The substrate used for this work is a n-type silicon wafer, doped with arsenic ($\rho = 0.01\,\Omega$cm) covered by an epitaxial layer with a thickness of 8 µm, doped with phosphorus ($\rho = 0.75\,\Omega$cm). The crystallographic orientation of the surface is in the (100) plane. The wafer used for the samples was cut in pieces of 1 cm x 1 cm with a thickness of 625 ± 25 µm.

2.1. Substrate cleaning

During the fabrication process, the contamination of particles on the surface of the wafer needs to be avoided in order to have a clean substrate for the further steps. A smooth and clean surface is necessary to deposit effectively a uniform graphenic carbon film. First, the sample was placed in acetone at 60 °C and sonicated for 10 min, in order to remove the protective resist layer. Next, the sample was put into isopropyl alcohol and sonicated with the same previously reported conditions to remove the acetone residuals not soluble in water. Afterward, the sample was prepared with a full RCA clean [10]. The standard cleaning SC-1 was done at 75 °C for 10 min to eliminate the organic residuals. After that, the standard cleaning SC-2 was performed at 75 °C for 10 min to remove ionic contaminations due to metallic particles on the surface.

2.2. Graphenic carbon deposition

Before the carbon deposition, it is necessary to dip the sample in 5% hydrofluoric acid (HF) for 150 sec to remove the native oxide. The deposition was performed with a low-pressure chemical vapor deposition (LPCVD) process at 1000 °C. The used precursor was C_2H_4 with a gas flow of 20 sccm and the pressure in the chamber was 20 mbar. This deposition was carried out for 7 min to obtain the desired layer of GC [6, 9].

2.3. Laser lithography

The lithography process is used to transfer the patterns to the substrate using a spin-coated negative photoresist and a UV laser. Firstly, the sample was heated on a hotplate up to 200 °C for 30 min to remove the adsorbed humidity and afterward, the resist ma-N 1420 was spin-coated on the cooled substrate at 3300 rpm for 30 sec. Next, the sample was heated for 90 sec at 100 °C to remove the solvents from the resist. In order to generate the desired structure, the surface of the sample was exposed to the focused laser beam of the *KLOÉ Dilase 250* laser writer with a laser wavelength of 375 nm and a power density of 351 kW cm^{-2}. Fig. 1 shows the process steps and the result after the laser lithography. The key parameters to control the design are the modulation of the laser intensity, the velocity of the stage, the focal plane, the exposure time and the possibility to apply three different optical filters. The minimum spot size of the laser is 1 µm but changing the focal plane a spot size around 100 µm has been obtained. The writing process requires only few minutes to produce more than hundreds of reproducible shapes using velocity and repositioning velocity of 2 mm s^{-1}. Finally, the sample was dipped in ma-D 533/s, which is a developer that removes the unexposed resist in around 1 minute. This developing time is particularly critical and can directly affect the geometric property and the resolution of the pattern.

2.4. Dry etching

The RF plasma process was used to remove the carbon from the area that is not covered by the resist. The pressure in the chamber was 2 mbar for a H_2 plasma and around 0.5 mbar for a O_2 plasma respectively. The effects of both plasmas were compared and evaluated under different conditions. The etching time was always set at 3-times 10 sec with 30 sec of cooling time. In the Fig. 2, the process steps and the result are shown. After the plasma etching, the resist was removed as reported in Sec. 2.1.

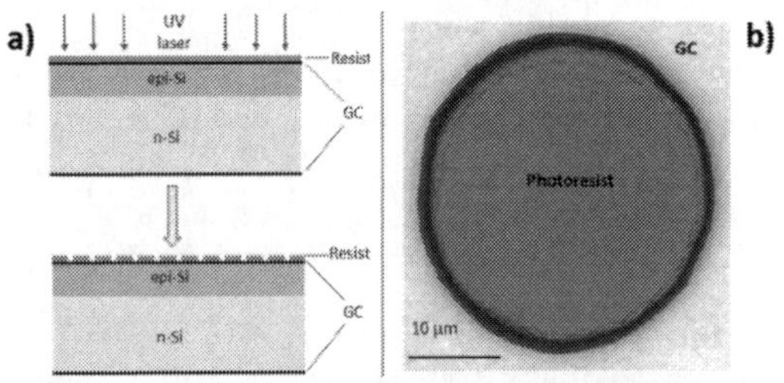

Figure 1. Certain areas of the surface are exposed to the focused laser beam to generate the desired pattern (a). An example of the exposed and developed circular resist pattern is shown in (b).

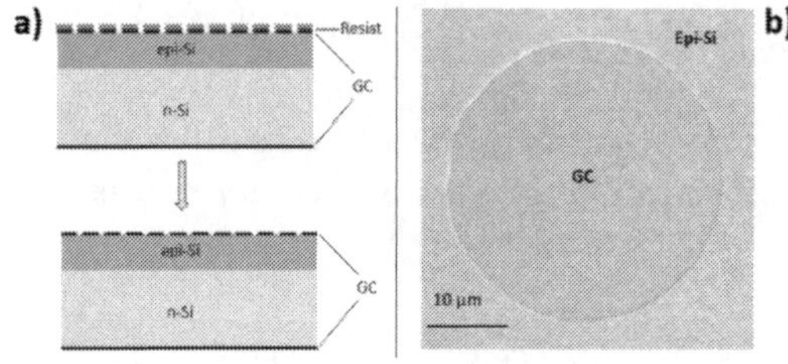

Figure 2. After the development of the resist, the pattern is created in the carbon by etching the carbon away. The process steps are illustrated in (a). (b) shows an optical true color image of a structure after removal of the resist.

2.5. Wet etching

Before being dipped into the KOH bath, the sample was immersed into HF for 30 sec to remove the native oxide that could delay the silicon etching. A 5 M solution of KOH saturated with isopropanol was heated up at 80 °C in a constant-temperature water bath and the sample was immersed for 9 min. After that, decontamination with the SC-2 cleaning was carried out for 30 min. The final result of this step and the relative schematic is reported in Fig. 3.

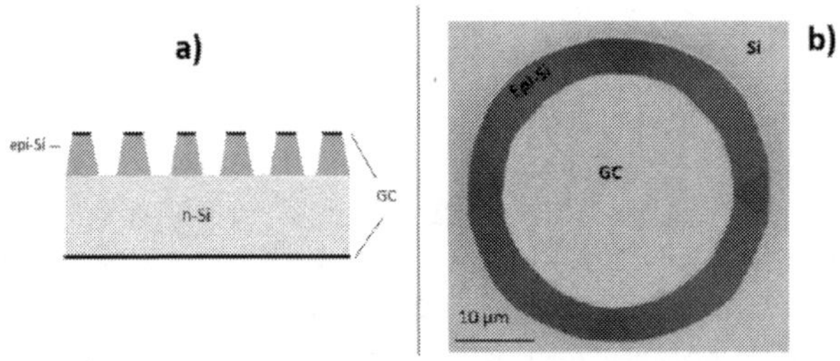

Figure 3. After the pattern of the carbon layer by plasma etching, the native oxide was removed by a HF bath and the silicon was etched away using KOH saturated with isopropanol. a) Shows a schematic of the wet etching result. b) (top view) 3D laser image of the circular structure after the wet etching process.

3. Results

In order to determine the presence of the GC on the substrate, Raman spectroscopy is used with a laser excitation at 532 nm. The D and G bands are the most relevant ones for the GC

IOP Conf. Series: Journal of Physics: Conf. Series **1226** (2019) 012011 doi:10.1088/1742-6596/1226/1/012011

spectrum and are shown in Fig. 4.

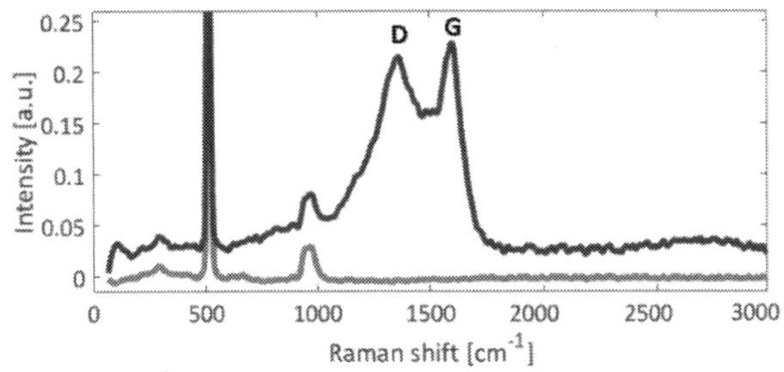

Figure 4. Raman spectra of GC after a deposition of 33 nm on silicon substrate (blue) and silicon without GC (red).

The Raman shift associated with the D and G peaks are at 1350 cm^{-1} and 1584 cm^{-1}. As reported by [11], the high content of covalent sp^2-bonds is highlighted by the height of the G peak compared to the D peak. After the plasma dry etching, the GC is removed and the two silicon peaks at 520 cm^{-1} and 960 cm^{-1} are clearly visible. After the GC deposition, as described in Sec. 2.2, the surface roughness of the film was measured with an atomic forced microscopy (AFM) set in tapping mode in order to not damage the surface. The root mean square height (Sq) is 1.3 ± 0.1 nm and the dashed white line in the Fig. 5 (a) shows the section where the profile of the film was evaluated. The thickness of the GC was measured with a 3D laser microscope (*Keyence VK series*) on a sample patterned with lines and crosses through the laser lithography and after the removal of the carbon from the area not covered by the resist by dry etching. In this way, a step to evaluate the height of the GC deposition was obtained. The dashed black line in the Fig. 6 (a) shows where the analysis was taken and the Fig. 6 (b) shows the step with the height of 33 ± 1 nm.

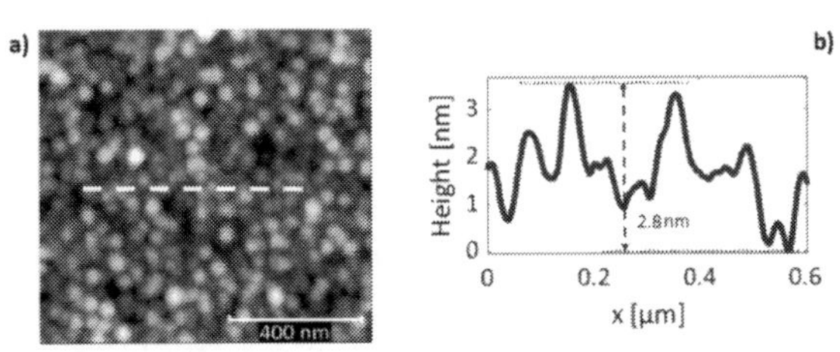

Figure 5. a) AFM image of a silicon sample after a deposition of GC. b) AFM profile of GC.

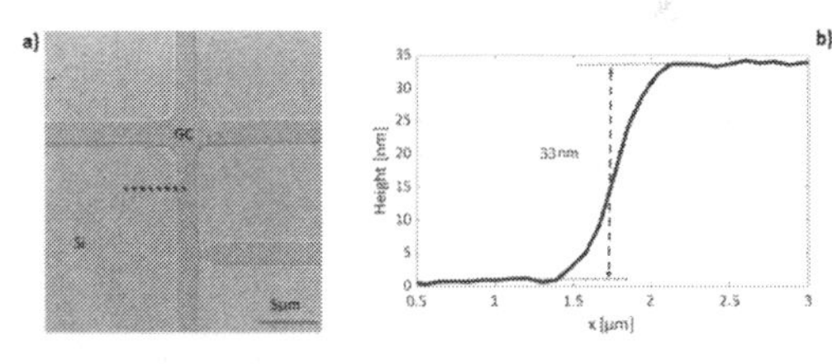

Figure 6. a) 3D laser microscope image of a GC cross. b) Cross profile to determine the thickness of the GC deposition.

Concerning the laser lithography, the resist used has a thickness of 2 µm using the spin coating parameters reported in Sec. 2.3. It was investigated how to obtain different diameters during the writing process by setting the focal plane, the exposure time and the modulation. In the plot (a) of the Fig. 7 the diameter function of the exposure time for different focal planes, at fixed modulation (80 %) and without a filter it is shown. In the plot (b) of Fig. 7 the exposure dose function of the exposure time at a different focal plane and at fixed modulation it is illustrated.

The value of the exposure dose D was evaluated according to the following equation:

$$D = \frac{P \cdot m}{(\pi r^2)} \cdot t \tag{1}$$

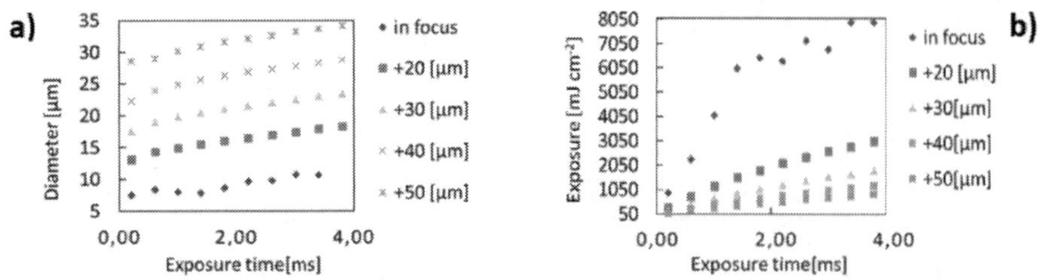

Figure 7. a) Experimental plot of the dot diameters function of the exposure time at fixed modulation and focal plane. b) Plot of the exposure dose transferred to the resist, function of the exposure time, at fixed focal plane and modulation. The relative error is $\pm 1.1\%$

where t is the exposure time, m is the laser power modulation (in %), r is the radius of the printed dot and P is derived from the maximum power density ($351\,\mathrm{kW\,cm^{-2}}$) given by the manufacturer and divided by the area of the laser spot with a diameter of $1\,\mu\mathrm{m}$ (resolution of the machine). This last calculation is only theoretical and was not yet been validated by measurements due to the complicated setup. However, the exposure dose required, as stated in the resist datasheet, is around $550\,\mathrm{mJ\,cm^{-2}}$ and it was always surpassed with the combination of the chosen parameter during the laser writing. Fig. 8, shows the effects of both plasmas on the resist mask. Mainly, due to the higher energy at a lower wavelength for the H_2 plasma (Lyman series), as depicted in Fig. 8 (b), the resist was hard baked and it was no longer possible to remove even after repeated cleaning in acetone and NMP. This is also confirmed by the absorbance coefficient of the resist, which shows a peak at a wavelength under 250 nm. In some cases, the resist hardbake occurred only at the boundary of the resist, due to the fact that this effect is strictly dependent on the previous process conditions modifying parameters such as the resist thickness. Therefore, the O_2 plasma was chosen for reliable and repeatable results as shown in Fig. 8 (a). The last evaluation was regarding the wet etching process. Using the *Keyence 3D laser scanner* as an optical profilometer, the side a and the angle α reported in the schematic in the Fig. 9 (a) were measured for each plane as shown in Fig. 9 (b).

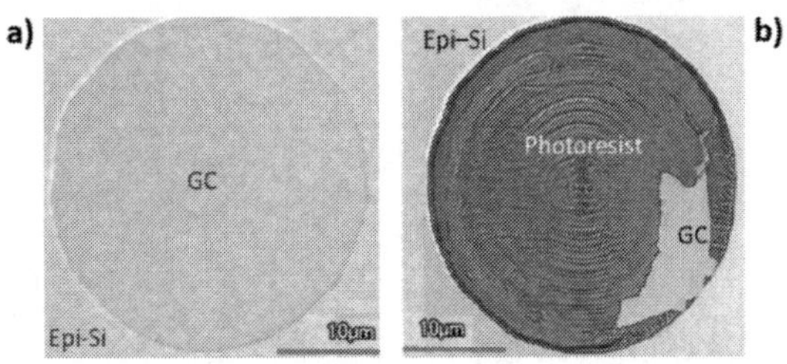

Figure 8. Comparison between two different plasmas during the dry etching process. a) effect of O_2 plasma on a circular resist pattern. Next, the resist over the GC layer was successfully removed by solvents. b) hardbake effect on the resist after H_2 plasma on a circular resist pattern.

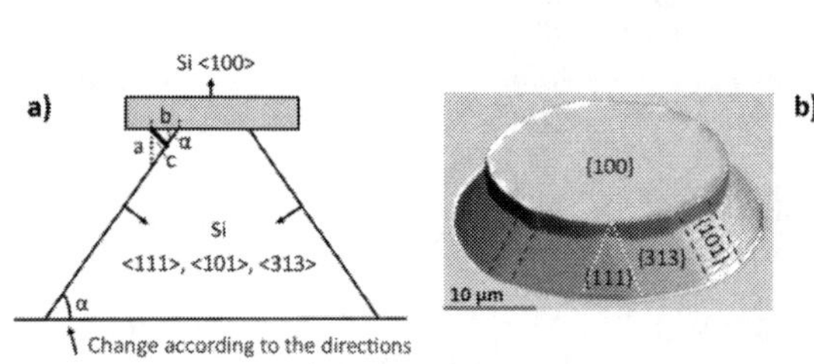

Figure 9. a) Schematic of the crystal plane section. b) 3D laser scanner image of a circle (top view) with a diameter of $31\,\mu\mathrm{m}$. The wet etching step was performed for 9 min and the absolute height obtained is $8.9\,\mu\mathrm{m}$. The investigated planes are highlighted, the other planes are symmetric [2].

Using the right-angled triangle definition:

$$c = a \cdot cos(\alpha) \tag{2}$$

and divide by the etching time the etch-rate for each crystal plane is estimated as reported in the Tab. 1.

Table 1. The etching rate (second row), expressed in $\mu m\,min^{-1}$, evaluated for each crystal plane (first row).

(100)	(111)	(313)	(101)
0.82	0.1	0.13	0.15

Fig. 10 shows a comparison between the use of KOH without isopropanol (a) and with isopropanol (b). In Fig. 10 (c) two problems are highlighted: the first concerning the effect of the H_2 plasma on the resist, as discussed before, and the second regarding the missing dip in HF before the KOH bath that produces hillocks on the substrate.

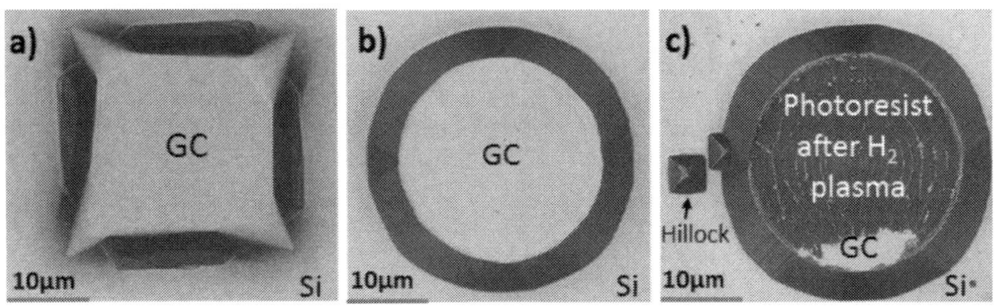

Figure 10. a) Effect of wet etching with (5M) KOH not saturated with isopropanol. b) Effect of wet etching: first HF(5 %) dip for 20 sec and after (5M) KOH saturated with isopropanol. c) Effects of H_2 plasma on the resist and (5M) KOH saturated with isopropanol without a prior dip in HF(5 %) and a resulting hillocks formation on the surface.

4. Conclusions

In conclusion, an easy method to pattern different shapes on (100) silicon substrates is shown. The polycrystalline GC as an etch mask for wet chemical processing offers different advantages such as excellent adhesion to silicon due to the formation of covalent bonds during the deposition, material thinness, resistance to common solvents, aggressive acids (HF) and bases (KOH), easy and quick removal by O_2 plasma. Furthermore, it is an excellent conductive material that could be directly used as a stable contact.

5. References

[1] French P J *et al.* 1997 *Sensors and Actuators A: Physical* **62**
[2] Barycka I *et al.* 1995 *Sensors and Actuators A: Physical* **48**
[3] Zubel I *et al.* 2004 *Sensors and Actuators A: Physical* **115**
[4] Zubel I *et al.* 2002 *Sensors and Actuators A: Physical* **101**
[5] Bianco A *et al.* 2013 *Carbon* **65**
[6] Kreupl F 2011 *MRS Proceedings* **1303**
[7] Huebner S *et al.* 2015 *IEEE Transactions on Nuclear Science* **62**
[8] Reinhard D K *et al.* 2004 *J. Vac. Sci. Technol. B: Microelectr. Nanom. Struct.* **22**
[9] Stelzer M *et al.* 2017 *IEEE Journal of the Electron Devices Society* **5**
[10] Kern W 1990 *Journal of the Electrochemical Society* **137**
[11] Pimenta M A *et al.* 2007 *Phys. Chem. Chem. Phys.* **9**

Second Magnetization Peak Effect in a Fe(Se,Te) iron based superconductor

A Galluzzi[1,2], K Buchkov[3], V Tomov[3], E Nazarova[3], A Leo[1,2], G Grimaldi[2], A Nigro[1,2], S Pace[1,2] and M Polichetti[1,2]

[1]Department of Physics "E.R. Caianiello", University of Salerno, via Giovanni Paolo II, 132, Fisciano (SALERNO), I-84084, Italy
[2]CNR-SPIN Salerno, via Giovanni Paolo II, 132, Fisciano (SALERNO), I-84084, Italy
[3]Georgy Nadjakov, Institute of Solid State Physics, Bulgarian Academy of Sciences, 72 Tzarigradsko Chaussee Blvd., 1784 Sofia, Bulgaria
E-mail: agalluzzi@unisa.it

Abstract. The iron based superconductor $FeSe_{0.5}Te_{0.5}$ has been investigated by means of DC magnetic measurements as a function of magnetic field (H). By considering the superconducting m(H) hysteresis loops at different temperatures, the sample shows a strong superconducting signal together with the presence of a peak effect that causes an anomalous increase in the field dependence of the critical current density $J_c(H)$. The presence of the peak effect has been studied by means of the $J_c(T)$ obtained at different magnetic fields starting from the $J_c(H)$ curves. The analysis of the $J_c(T)$ curves shows that the peak effect is due to a crossover from a weak pinning regime to a strong pinning regime.

1. Introduction

The discovery of the iron based superconductors (IBS) in 2008 [1] and in particular the analysis of their fascinating superconducting properties [1–3] gave a boost to the researchers interest for taking into account the possibility of using these materials for applications. Among the IBS, the Fe(Se,Te) is one of the most attractive compound due to its chemical stability, pinning capabilities and rich vortex dynamics. In the framework of the vortex phenomena, one of the most intriguing property is the second magnetization peak effect (named also peak effect) that consists in a non-monotonous behavior of the field dependence of the critical current density $J_c(H)$. This particular $J_c(H)$ trend can be ascribed to several reasons such as matching effect, surface barriers, dynamic effects, different types of phase transition etc. In the years this phenomenon has been widely investigated in Fe(Se,Te) both single crystals and bulk samples [4–6]. The nature of the peak effect phenomenon is very complicated because of the several parameters that can be involved in the birth and the evolution of it such as the material structure, the samples morphology or also pinning energy changes.

In this work we want to study the possible cause of the presence of a peak effect found by means of dc magnetic measurements as a function of the field. At this aim we have performed m(H) measurements at different temperatures and calculated the field dependence of the critical current density $J_c(H)$ that have shown the clear presence of the peak effect. Starting from these curves, we have extracted the temperature dependence of the critical current density $J_c(T)$ considering a large interval of field. By fitting the $J_c(T)$ curves with different pinning models, the cause of the peak effect birth has been determined.

2. Experimental details

We have analyzed a Fe(Se,Te) sample with dimensions 3 x 3 x 0.2 mm^3 derivative of the basic FeSe [7]. The sample fabrication details are reported elsewhere [8]. The sample has been characterized using a Quantum Design PPMS-9T equipped with a VSM option. The dc magnetic measurements have been performed as a function of the magnetic field m(H) applying it perpendicular to the sample largest face. To avoid the effect on the sample response due to the residual field inside the superconducting dc magnet, the entrapped field was zeroed before each measurement [9]. For the m(H) measurements, the sample was first cooled down to the measurement temperature in absence of field (Zero Field Cooling) and thermally stabilized for at least 20 minutes. Then, the field was ramped with a sweep rate of 0.01 T/s to reach +9 T, then back to -9 T, and finally to +9 T again in order to acquire the complete M(H) loop.

3. Results and discussion

In order to study the transport and pinning properties of the sample, measurements of magnetic moment (m) as a function of magnetic field (H) have been performed at different temperatures as reported in Figure 1 in the case of positive fields. Looking the m(H) curves, it can be noted a strong superconducting signal accompanied by the presence of a second magnetization peak effect (or only peak effect) that causes a non-monotonous behavior of the curves. A shift of the second peak in the m(H) curves is evident and in particular its position moves to lower fields as the temperature increases. A symmetric behavior is obtained for negative values of magnetic field. It was not possible to individuate the peak effect for T > 10 K, probably due to the competition between the small superconducting signal and a non-zero magnetic signal near T_c originating from magnetic impurities present in our sample.

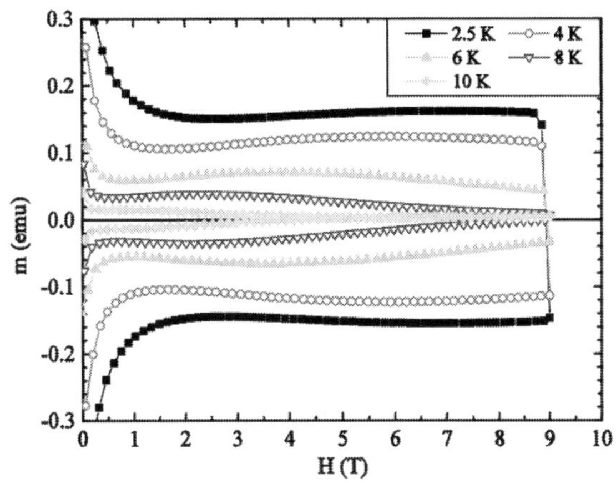

Figure 1. Field dependence of magnetic moment for different temperatures. The non-monotonous behavior of the curves is well visible.

In order to analyze the peak effect, the field and temperature dependencies of the critical current density $J_c(H,T)$ have been considered. In particular, from the m(H) curves of Figure 1, the critical current densities $J_c(H)$ have been obtained at different temperatures by means of the Bean critical state model[10,11]

$$J_c = \frac{20\Delta M}{\left[a\left(1 - \frac{a}{3b}\right)\right]} \quad (1)$$

where $\Delta M = M_{dn} - M_{up}$ are the magnetization measured with decreasing and increasing applied field, respectively, and a and b are the lengths (in cm) characterizing the cross section of the sample perpendicular to the applied field (H||c). The unit of ΔM is in electromagnetic unit per centimeter cubic and the calculated J_c is in Ampere per square centimeter. The obtained $J_c(H)$ curves are reported in Figure 2 for different temperatures. In particular, the $J_c(H)$ curves have been reported in a log-log scale that allows to observe the low dependence of J_c as a function of the magnetic field that is a clue of the strength of the pinning inside the material. In addition, the peak effect characterized by an anomalous increase of the J_c as a function of the field is well visible. By carrying out isofield cuts of J_c for certain fixed fields from the $J_c(H)$ curves at different temperatures, the $J_c(T)$ dependencies have been extracted in order to obtain information about the pinning regime acting in the sample. At this aim, the $J_c(T)$ data have been analyzed and fitted with several pinning models reported in the literature. In particular, the first two fits were performed considering the δl pinning and the δT_c pinning [12]. The first one is due to the spatial variation in the charge carrier mean free path while the second one is due to randomly distributed spatial variation in T_c. According to the theoretical approach proposed by Griessen [13], the δl pinning is described by the equation $J_c(t)=J_c(0)(1-t^2)^{5/2}(1+t^2)^{-1/2}$ while the δT_c pinning is described by $J_c(t)=J_c(0)(1-t^2)^{7/6}(1+t^2)^{5/6}$, where $J_c(0)$ is the value of J_c at T = 0 K and $t = T/T_c$. These equations have been used to fit the J_c(T) curve for a low field value, exactly H = 0.5 T. As shown in the main panel of Figure 3, it is clear that neither the δl pinning (red solid line) nor the δT_c pinning (green dashed line) fit the data.

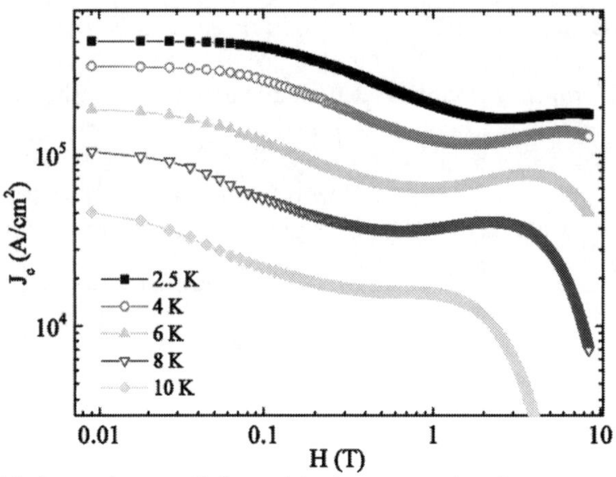

Figure 2. Magnetic field dependence of the critical current density J_c reported in log-log scale at different temperatures extracted from the M(H) curves of Figure 1.

After that, a model assuming the Ginzburg-Landau temperature dependencies for the thermodynamical critical magnetic field [14] has been tried to fit the data using the equation $J_c(t) = J_c(0)(1-t^2)(1+t^2)$ where $J_c(0)$ is the value of J_c at T = 0 K and $t = T/T_c$. The blue dotted line of Figure 3 shows that also in this case the fit is not satisfactory. Another fit was performed considering a model that takes into account the existence of giant flux creep [15] and using the equation $J_c(t)=J_c(0)(1-t^2)^2$ where $J_c(0)$ is the value of J_c at T = 0 K and $t = T/T_c$. The cyan dashed-dotted line associated to the giant flux creep does not reproduce the data. Finally, by following the approach of Polat et al [16], we have considered equations that predict the presence of weak or strong pinning centers inside the sample. For what concerns the weak pinning, the equation describing the critical current in presence of this type of pinning centers, in particular considering lattice defects associated to point-like disorder or to in-plane dislocations inside the sample[12,16], is $J_c^{weak}(\mathrm{T}) = J_c^{weak}(0)\, e^{-T/T_0}$ where $J_c^{weak}(0)$ is the value of J_c at T = 0 K, and T_0 is the characteristic pinning energy of weak (typically point-like) pinning defects. For what regards the strong pinning, the equation describing the critical current in presence of this type of pinning centers, in particular considering defects generated by correlated disorder such as twin boundaries, columnar pins or defects[16–18], is $J_c^{str}(\mathrm{T}) = J_c^{str}(0)\, e^{-3(T/T^*)^2}$ where $J_c^{str}(0)$ is the value of J_c at T = 0 K, and T* is the vortex pinning energy of strong pinning centers. Fitting the data of Figure 3 with the strong pinning expression (purple short dashed line), it can be noted that the fit is better than the other pinning models used so far but it presents big deviations from the data at certain temperatures. On the other hand, using the weak pinning expression (see inset of Figure 3), the fit describes very well the data. This has been verified for all the magnetic fields minor than 1 Tesla. So, in this low field region (H ranging from 0.1 T to 0.9 T) the best fit was obtained by using the equation describing the critical current in presence of weak pinning centers.

Now we want to analyze the J_c(T) curves behavior considering a high magnetic field value, for example H = 7 Tesla, in order to understand if the pinning regime acting in the sample is still characterized by the weak pinning centers. Following the previous approach, it is worth to underline that δl pinning and the δT_c pinning are effective only in the single vortex pinning regime, i.e. in low-field and zero-field regions [19,20]. For this reason they have been excluded in the analysis of the curves at high fields.

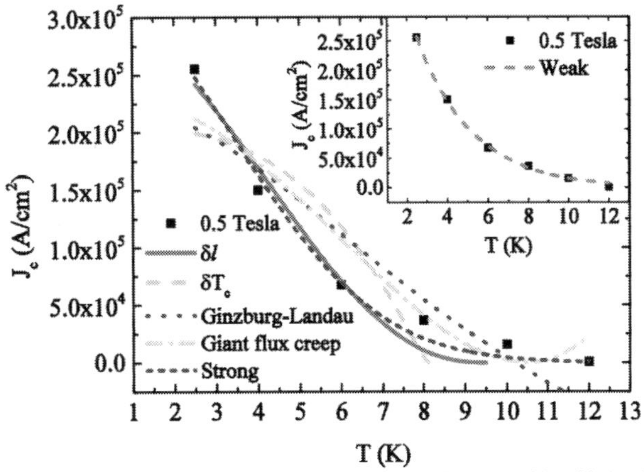

Figure 3. Temperature dependence of J_c at H = 0.5 Tesla fitted with δl (red solid line), δT_c (green dashed line), Ginzburg-Landau (blue dotted line), giant flux creep (cyan dashed-dotted line) and strong (purple short dashed line) pinning model. Inset: $J_c(T)$ at H = 0.5 Tesla fitted with weak pinning model (magenta dashed line).

In Figure 4 it is evident that the Ginzburg-Landau (blue dotted line) together with the weak (magenta dashed line) pinning model are the worst models for describing the data at 7 T. In addition, the giant flux creep model (cyan dashed-dotted line) characterizes well the data at 7 T but it predicts an increase of the J_c values at high temperatures that is not observed in our case and for this reason it has to be discarded. In the inset of Figure 4, it is shown the fit of the data at 7 T with the strong pinning model (purple short dashed line) and it is clear that the fit in this case is good and describes very well the experimental data. This has been verified for all the magnetic fields H > 3 T. So, in the high field region (H ranging from 3 T to 8.5 T) the best fit was obtained by using the equation describing the critical current in presence of strong pinning centers. At this point it is reasonable to think that in the intermediate region 1 T ≤ H < 3 T the $J_c(T)$ can be described by a combination of the weak and the strong pinning contributions (fit not reported), expressed through the equation [16] $J_c(T) = J_c^{weak}(0)\, e^{-T/T_0} + J_c^{str}(0)\, e^{-3(T/T^*)^2}$ where $J_c^{weak}(0)$, $J_c^{str}(0)$, T_0 and T* are the same parameters seen previously for weak and strong expressions.

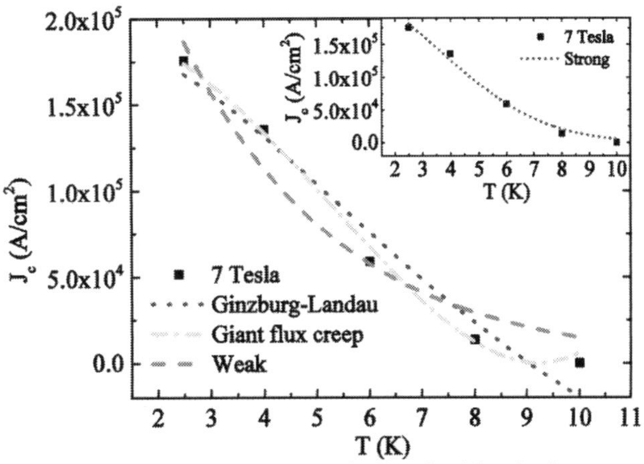

Figure 4. Temperature dependence of J_c at H = 7 Tesla fitted with Ginzburg-Landau (blue dotted line), giant flux creep (cyan dashed-dotted line) and weak (magenta dashed line) pinning model. Inset: $J_c(T)$ at H = 7 Tesla fitted with strong pinning model (purple short dashed line).

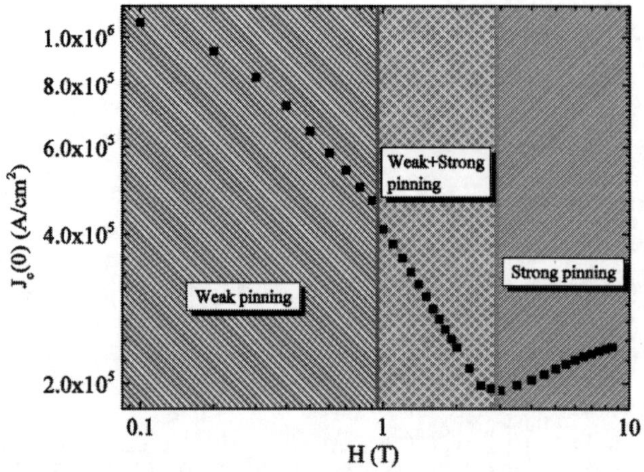

Figure 5. $J_c(T = 0\ K)$ versus H. The crossover among different pinning regimes, which determines the beginning of the peak effect phenomenon, is visible.

From the fittings by weak, strong and weak+strong equations it is possible to obtain the values for the parameters $J_c^{weak}(0)$, $J_c^{str}(0)$ and $J_c^{weak}(0) + J_c^{str}(0)$, respectively, i.e. the zero-temperature critical current densities $J_c(0)$ as a function of the magnetic field. These values are plotted in a log-log scale in Figure 5. It is clear that the critical current density has a non-monotonous behavior also at T = 0 K and in particular its value starts to increase when the weak pinning centers are ineffective and the contribution to the pinning is given only by the strong pinning centers. On the basis of the reported results we can claim that the crossover between weak and strong pinning regimes determines the birth of the peak effect phenomenon of the sample. A summary of the pinning models used for fitting the $J_c(T)$ data in low and high field region is reported in Table I.

Table I. *Fit of the data with the main pinning models in low and high field regions.*

Pinning model	Low field region (H < 1 T) Data fitted?	High field region (H > 3 T) Data fitted?
δl	No	Not possible
δT_c	No	Not possible
Ginzburg-Landau	No	No
Giant flux creep	No	No
Weak	Yes	No
Strong	No	Yes

4. Conclusions

By means of DC magnetic measurements as a function of magnetic field (H) we have analyzed the magnetic behavior of a $FeSe_{0.5}Te_{0.5}$ sample. In particular, performing m(H) measurements at different temperatures we have observed wide superconducting hysteresis loops accompanied by the presence of a peak effect. This phenomenon is visible in the anomalous increase of the field dependence of the critical current density $J_c(H)$ extracted from the m(H) loops at different temperatures. From these curves, by performing isofield cuts of J_c at fixed fields, the $J_c(T)$ curves have been acquired. Fitting the $J_c(T)$ dependencies with several pinning models at different magnetic fields ranging from 0.1 T up to 8.5 T, the information about the pinning regime acting in the sample have been obtained. In particular, it has been found that the sample undergoes a pinning crossover from a weak pinning regime acting until H > 1 T, to a strong pinning regime when H > 3 T. Observing the J_c (T = 0 K) values as a function of field, obtained from the $J_c(T)$ fits, it is evident that the origin of the peak effect can be ascribed to this weak/strong pinning crossover.

References

[1] Kamihara Y, Watanabe T, Hirano M and Hosono H 2008 Iron-based layered superconductor La[O(1-x)F(x)]FeAs (x = 0.05-0.12) with T(c) = 26 K. *J. Am. Chem. Soc.* **130** 3296–7
[2] Leo A, Braccini V, Bellingeri E, Ferdeghini C, Galluzzi A, Polichetti M, Nigro A, Pace S and Grimaldi G 2018 Anisotropy Effects on the Quenching Current of Fe(Se,Te) Thin Films *IEEE Trans. Appl. Supercond.* **28** 1–4
[3] Grimaldi G, Leo A, Nigro A, Pace S, Braccini V, Bellingeri E and Ferdeghini C 2018 Angular dependence of vortex instability in a layered superconductor: the case study of Fe(Se,Te) material *Sci. Rep.* **8** 4150
[4] Miu D, Noji T, Adachi T, Koike Y and Miu L 2012 On the nature of the second magnetization peak in FeSe1− xTex single crystals *Supercond. Sci. Technol.* **25** 115009
[5] Galluzzi A, Buchkov K, Tomov V, Nazarova E, Leo A, Grimaldi G, Nigro A, Pace S and Polichetti M 2018 Evidence of pinning crossover and the role of twin boundaries in the peak effect in FeSeTe iron based superconductor *Supercond. Sci. Technol.* **31** 015014
[6] Bonura M, Giannini E, Viennois R and Senatore C 2012 Temperature and time scaling of the peak-effect vortex configuration in FeTe0.7Se0.3 *Phys. Rev. B* **85** 134532
[7] Fiamozzi Zignani C, Corato V, Leo A, De Marzi G, Mancini A, Takano Y, Yamashita A, Polichetti M, Galluzzi A, Rufoloni A, Grimaldi G and Pace S 2016 Fabrication and Characterization of Sintered Iron-Chalcogenide Superconductors *IEEE Trans. Appl. Supercond.* **26** 1–5
[8] Galluzzi A, Buchkov K, Tomov V, Nazarova E, Kovacheva D, Leo A, Grimaldi G, Pace S and Polichetti M 2018 Mixed state properties of iron based Fe(Se,Te) superconductor fabricated by Bridgman and by self-flux methods *J. Appl. Phys.* **123** 233904
[9] Galluzzi A, Mancusi D, Cirillo C, Attanasio C, Pace S and Polichetti M 2018 Determination of the Transition Temperature of a Weak Ferromagnetic Thin Film by Means of an Evolution of the Method Based on the Arrott Plots *J. Supercond. Nov. Magn.* **31** 1127–32
[10] Bean C P 1962 Magnetization of hard superconductors *Phys. Rev. Lett.* **8** 250–3
[11] Bean C P 1964 Magnetization of High-Field Superconductors *Rev. Mod. Phys.* **36** 31–9
[12] Blatter G, Feigel'Man M V., Geshkenbein V B, Larkin A I and Vinokur V M 1994 Vortices in high-temperature superconductors *Rev. Mod. Phys.* **66** 1125–388
[13] Griessen R, Wen Hai-hu, van Dalen A J J, Dam B, Rector J and Schnack H G 1994 Evidence for mean free path fluctuation induced pinning in YBa2Cu3O7 and YBa2Cu4O8 films *Phys. Rev. Lett.* **72** 1910–3
[14] Di Gioacchino D, Celani F, Tripodi P, Testa A M and Pace S 1999 Nonuniversal temperature dependencies of the low-frequency ac magnetic susceptibility in high-Tc superconductors *Phys. Rev. B* **59** 11539–50
[15] Yeshurun Y and Malozemoff a. P 1988 Giant flux creep and irreversibility in an Y-Ba-Cu-O crystal: An alternative to the superconducting-glass model *Phys. Rev. Lett.* **60** 2202–5
[16] Polat A, Sinclair J W, Zuev Y L, Thompson J R, Christen D K, Cook S W, Kumar D, Chen Y and Selvamanickam V 2011 Thickness dependence of magnetic relaxation and E-J characteristics in superconducting (Gd-Y)-Ba-Cu-O films with strong vortex pinning *Phys. Rev. B - Condens. Matter Mater. Phys.* **84** 024519
[17] Nelson D R and Vinokur V M 1993 Boson localization and correlated pinning of superconducting vortex arrays *Phys. Rev. B* **48** 13060–97
[18] Hwa T, Le Doussal P, Nelson D R and Vinokur V M 1993 Flux pinning and forced vortex entanglement by splayed columnar defects *Phys. Rev. Lett.* **71** 3545–8
[19] Ghorbani S R, Wang X L, Hossain M S A, Dou S X and Lee S-I 2010 Coexistence of the δl and δTc flux pinning mechanisms in nano-Si-doped MgB 2 *Supercond. Sci. Technol.* **23** 025019
[20] Lei H, Wang K, Hu R, Ryu H, Abeykoon M, Bozin E S and Petrovic C 2012 Iron chalcogenide superconductors at high magnetic fields *Sci. Technol. Adv. Mater.* **13** 054305

IOP Conf. Series: Journal of Physics: Conf. Series **1226** (2019) 012013 doi:10.1088/1742-6596/1226/1/012013

Space charge limited current and photoconductive effect in few-layer MoS_2

A Grillo[1], F Giubileo[2], L Iemmo[2,3], G Luongo[1,2], F Urban[1,2], and A Di Bartolomeo[4,5]

[1] PhD student, Physics Department "E. R. Caianiello", University of Salerno, via Giovanni Paolo II 132, Fisciano 84084, Salerno, Italy
[2] Researcher, CNR-SPIN, via Giovanni Paolo II 132, Fisciano 84084, Salerno, Italy
[3] Researcher, Physics Department "E. R. Caianiello", University of Salerno, via Giovanni Paolo II 132, Fisciano 84084, Salerno, Italy
[4] Associate Professor, Physics Department "E. R. Caianiello", University of Salerno, via Giovanni Paolo II 132, Fisciano 84084, Salerno, Italy.

[5] Corresponding author's email: adibartolomeo@unisa.it

Abstract. We analyse the conduction mechanism and the electrical photoresponse of chemical-vapor deposited MoS_2 few-layers on SiO_2/Si substrate. We perform temperature dependent I-V measurements and report a space-charge limited conduction due to the presence of an exponential distribution of trap states in the MoS_2 band-gap. We estimate the density of trap states as $10^{10} - 10^{11}\ cm^{-2}$ from the temperature-independent critical drain-source voltage. We also investigate the MoS_2 photocurrent under white light at different incident powers. We use a modified Hornbeck-Hayens model to study the photoconductive effect and for an alternative estimation of the trap state density.

1. Introduction

During the last few years, molybdenum disulfide (MoS_2), a transition metal dichalcogenide, has attracted a lot of attention as one of the most promising layered materials for new advanced electronic and optoelectronic devices[1-3]. Nowadays large-scale high-crystalline MoS_2 flakes, up to the centimeter size, are produced by chemical vapor deposition (CVD) from solid state precursors[4]. In a MoS_2 single layer, Mo (+4) and S (-2) atoms are arranged in a S-Mo-S sandwich structure by covalent bonds; single layers, held together by relatively weak van der Waals forces, form multi-layers flakes[5]. In its bulk form, MoS_2 is an indirect semiconductor with bandgap of 1.2–1.3 eV, while monolayer MoS_2 becomes a direct semiconductor with bandgap increasing up to 1.8–1.9 eV[6,7]. Differently from graphene[8,9], the wide bandgap constitutes a remarkable advantage of this material, enabling for instance selective light absorption in optoelectronic applications or high-performance switching in field effect transistors (FETs)[10,11]. In this work, we analyze the temperature dependent current-voltage (I-V) characteristics of CVD synthesized MoS_2 few-layer flakes to shed light on the mechanisms of conduction. We find non-linear characteristics, where the current depends on a certain power of the applied voltage, explained with a model based on the trap-induced space-charge limited conduction (SCLC). Then, we study the photocurrent of the MoS_2 few-layers under white LED illumination at different intensities. We focus on the photoconductive effect, which is the increase in conductivity due to photo-generation of electron-hole pairs, facilitated by MoS_2 band tail trap states, caused by intrinsic defects or MoS_2/SiO_2 interaction.

2. Experimental

MoS_2 flakes were synthesized by CVD on 300 nm SiO_2 thermal oxide grown over a (100) heavily doped p-type Si substrate (Fig.1a). The growth process occurred from sulfur (S) and molybdenum trioxide (MoO_3) powders at a temperature of 850 °C for 10 min under 50 sccm Ar gas flow. A flake, consisting of few layers, was selected and contacted with Ti/Au (15/120 nm) metals by standard electron-beam lithography (EBL) and lift-off process (Fig.1b). In the following, we focus on two-probe I-V measurements carried out between leads 1 and 2 of Fig. 1b.

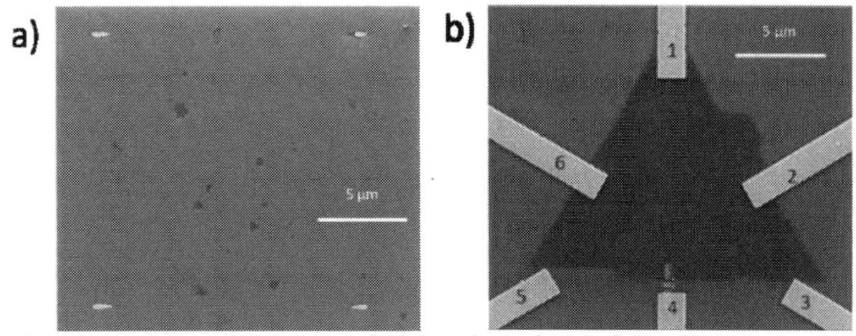

Fig. 1 - a) Scanning electron microscope (SEM) image of CVD synthesized MoS_2 flakes. The red arrow indicates the flake chosen for the measurements. b) SEM image of the selected few-layer MoS_2 flake after Ti/Au deposition. Three layers are clearly distinguishable in the SEM image.

Electrical measurements were performed with a semiconductor characterization analyzer Keithley 4200 SCS connected to a Janis ST-500 cryogenic probe-station with temperature (from 5K to 450K) and pressure (from 10^{-6} mbar to ambient) control. Photoconduction was investigated using the white light from a LED array with tunable optical power up to 5.5 $mWcm^{-2}$ and spectrum over the range 420-720 nm with peaks at 450 and 540 nm.

3. Results and discussion

3.1. Temperature dependent I-V characteristics

It is well-known that the distribution of trap states can be derived from current-voltage measurements at different temperatures, when the conduction is space-charge limited. This technique has been widely used to investigate the properties of trap states in low-mobility devices[13-15]. The space-charge limited conduction occurs when the injected carrier density exceeds the intrinsic carrier density of the material at a given temperature T. According to the SCLC theory, in the presence of an exponential charge trap distribution, $h(E) = \frac{P_t}{k_B T_c}\exp(-\frac{E}{k_B T_c})$, the current-voltage relationship is[14]:

$$J = \frac{\mu N_c}{q^{m-2}}\left(\frac{2m-1}{m}\right)^m \left(\frac{\varepsilon_0 \varepsilon_r (m-1)}{P_t m}\right)^{m+1} \left(\frac{V^m}{L^{2m-1}}\right) \quad (1)$$

Where P_t is the total trap density, T_c is a characteristic temperature of the trap distribution, k_B is the Boltzmann constant, N_c is the effective density of states in conduction band, E is the energy, q is the electric charge, ε_0 and ε_r are the vacuum and the relative permittivity, $m = \frac{T_c}{T} + 1$ and L is the distance between the two electrodes. From the definition of m, it is evident that the current is strongly dependent on the temperature and that m decreases with increasing T, until the minimum value of one. Fig. 2a shows the I-V characteristics of the device between leads 1 and 2, used as the source and the drain of the three-terminal structure (the third terminal, constituted by the Si back-gate, was grounded), at several temperatures in the range 117 K to 327 K. The temperature range is obtained by blowing liquid nitrogen into the chamber and it has been chosen to avoid damaging the sample.

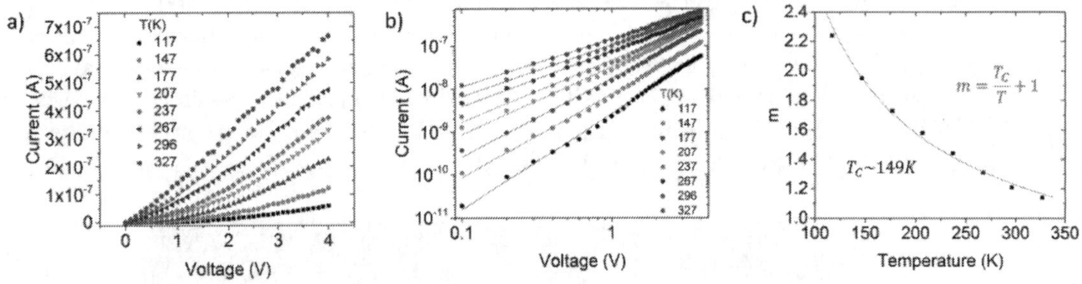

Fig. 2 - a) Two probe temperature dependent I-V characteristics of the MoS_2 few-layer flake. b) Linear fit of the two-probe I-V characteristics of the MoS_2 few-layer flake on a log-log scale and at different temperatures. c) Values of m, resulting from the linear fits in (b), plotted as function of the temperature T.

According to eq. (1), $J{\sim}V^m$; by plotting the I-V data on a log-log graph and carrying out a linear fit for each temperature, we are able to check the relation between m and the temperature. As shown in Fig.2c, the dependence of m on T is consistent with the SCLC theory, with $T_c \sim 149K$. This finding provides confirmation of the presence of exponentially distributed tail trap states in the MoS_2 bandgap. Therefore, we can exploit one of the main aspects of the trap-induced limited space charge conduction, i.e. that the density of trap states can be quantitatively determined from the measured I-V-T characteristics. As the applied voltage increases, the trap states get filled up by injected charge carriers until a critical voltage is reached, for which the conductivity of the sample becomes independent of T and the I-V characteristics at different temperatures intersect each other. The relation between this critical voltage V_c and the density of trap states P_t is given by[16]:

$$V_c = \frac{qP_tL^2}{2\varepsilon_0\varepsilon_r} \qquad (2)$$

Fig. 3 shows the extrapolation from the I-V curves (of Fig.2b) at different temperatures, with intersection at $V_c{\sim}13$ V, which according to eq. (2) corresponds to $P_t{\sim}10^{10}\ cm^{-2}$. The result is in good agreement with values quoted in the literature for similar devices. Such trap density corresponds to a MoS_2 flake with a medium level of defects[18,19,20].

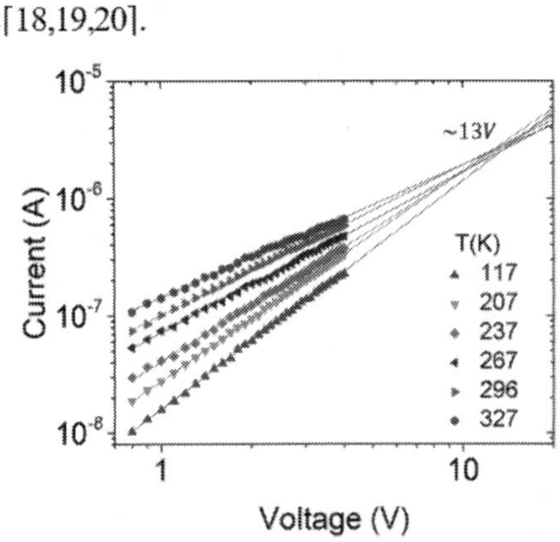

Fig. 3 – Extrapolation of the temperature independent critical drain-source voltage.

3.2. Photoconduction

IOP Conf. Series: Journal of Physics: Conf. Series **1226** (2019) 012013 doi:10.1088/1742-6596/1226/1/012013

Here we investigate the electrical photo-response of the MoS_2 few-layer flake. Fig. 4 shows I-V measurements under white LED illumination at different intensities.

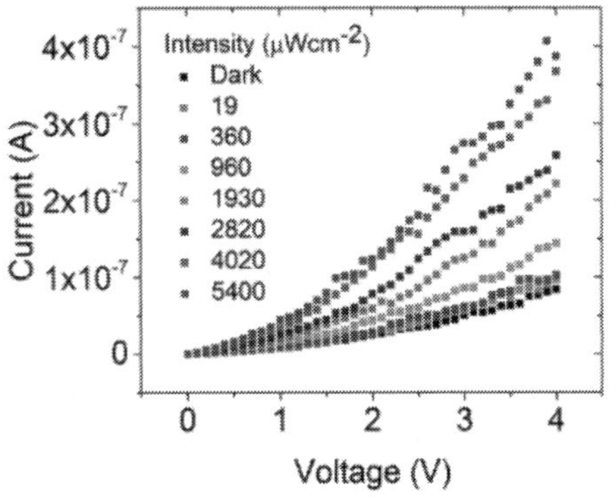

Fig. 4 –I-V characteristics under white LEDs illumination at different intensities.

There are two different mechanisms that can originate photoconductivity in the flake over the SiO_2/Si substrate: the photogating (PG) and the photoconductive (PC) effect[1,17]. The total photocurrent, then, is the sum of these two contributions:

$$I_{ph} = I_{PG} + I_{PC} \tag{3}$$

The PG effect manifest itself as a shift of the transfer characteristic of the transistor, or equivalently as a change in threshold voltage, resulting from the use of the Si substrate as the back-gate. Since we did not operate the device as a transistor (all measurements were performed at grounded back-gate), we can neglect it.
The PC effect, on the other hand, produces an increment in conductivity due to photo-induced excess carriers, and is facilitated by the presence of gap trap states:

$$I_{PC} = \left(\frac{W}{L}\right) V_D \Delta\sigma \tag{4}$$

where W and L are the width and the length of the conductive channel (here roughly assumed as L=6 μm and W=1.5 μm, given the irregular channel geometry), V_D is the applied bias, and $\Delta\sigma$ is the variation of the conductivity caused by light.
To evaluate $\Delta\sigma$, we use a modified Hornbeck-Hayens model[18,21] in which we approximate the exponential distribution of trap states in the band-gap by a narrow square distribution of states with density P_t (we consider only hole trapping and we neglect electron trapping because the n-type MoS_2 transistor operated above threshold – see following - has Fermi level situated near the conduction band edge so that most of the electron traps are filled). Under optical illumination, charge neutrality must be respected, so we can write:

$$\Delta n = \Delta p + p_t \tag{5}$$

where p_t is the trapped hole carrier concentration, and Δn and Δp denote the excess free electron and hole concentrations, respectively. Hence, the excess conductivity is:

$$\Delta\sigma = q(\mu_n + \mu_p)\Delta p + q\mu_n p_t \tag{6}$$

We remark that the photo-conductivity is higher compared to a scenario without traps by the amount $q\mu_n p_t$. The differential equations that describe the carrier dynamics are:

$$\frac{d\Delta p}{dt} = \varphi - \frac{\Delta p}{\tau_r} + \frac{p_t}{\tau_g} - \frac{\Delta p}{\tau_t}\left(1 - \frac{p_t}{P_t}\right) \tag{7}$$

$$\frac{dp_t}{dt} = -\frac{p_t}{\tau_g} + \frac{\Delta p}{\tau_t}\left(1 - \frac{p_t}{P_t}\right) \tag{8}$$

Where τ_t^{-1} and τ_g^{-1} are respectively the hole trapping and escape rates, $\tau_r \approx 100\ ps$ is the nonradiative carrier lifetime[22], φ is the flux of photons and P_t is the density of trap states. Imposing the steady-state conditions $\frac{d\Delta p}{dt} = \frac{dp_t}{dt} = 0$, we obtain:

$$\Delta p = \varphi \tau_r \tag{9}$$

$$p_t = \frac{\varphi P_t \tau_r}{\varphi \tau_r + P_t(\frac{\tau_t}{\tau_g})} \tag{10}$$

Considering low injection levels, eq. (10) can be reduced to:

$$p_t = \varphi \tau_r \left(\frac{\tau_g}{\tau_t}\right) \tag{11}$$

Inserting equations (6), (9), (10) and (11) into the relation for I_{PC}, eq. (4), we can obtain

$$I_{PC} = \left(\frac{W}{L}\right) V_D [q(\mu_n + \mu_p)\varphi \tau_r + \frac{q\mu_n \varphi P_t \tau_r}{\varphi \tau_r + P_t\left(\frac{\tau_t}{\tau_g}\right)}] \tag{12}$$

$$I_{PC} = \left(\frac{W}{L}\right) V_D [q(\mu_n + \mu_p)\varphi \tau_r + q\mu_n \varphi \tau_r \left(\frac{\tau_g}{\tau_t}\right)] \tag{13}$$

for the high and low injection regimes, respectively. We use eq. (12) to estimate the density of trap states P_t. To do that, we first obtain the ratio $\frac{\tau_g}{\tau_t}$ for each light intensity from the fitting of eq. (13) to the photocurrent at low injection, i.e. at low V_D (Fig. 5a). Then we use these values in eq. (12) to obtain the density of trap states P_t for each light intensity (Fig. 5b), which averaged, yield the final P_t estimation. The mobility needed to perform the first step is estimated from the transfer characteristic of the transistor, shown in the inset of Fig. 6b. We assume that electrons and holes have similar mobilities, $\mu_n \approx \mu_p = \mu$, and calculate it from the usual relation $\mu = \frac{L}{W C_{ox} V_{ds}} \frac{dI_{ds}}{dV_{gs}}$ [1,10], obtaining $\mu_n = \mu_p = 0.6 \frac{cm^2}{Vs}$.

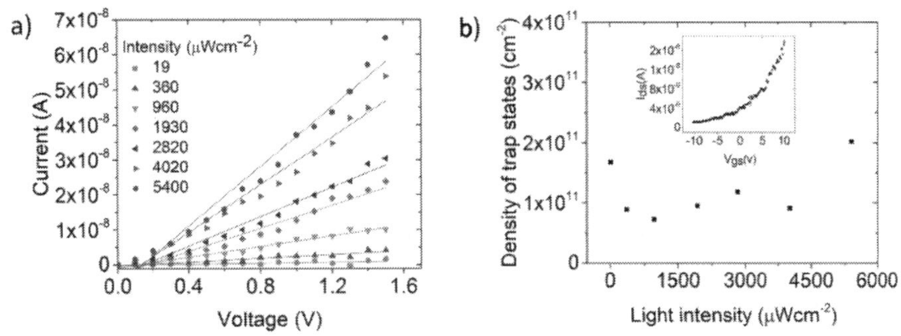

Fig. 5 – a) Linear fit of the photocurrent plotted as function of the applied voltage b) MoS_2 density of trap states as function of the intensity of the incident light. Inset: transfer characteristic of the MoS_2 flake.

The result of the above described procedure is $P_t \sim 10^{11} cm^{-2}$, which considering the approximations and the several parameters involved, can be considered consistent with the result obtained from the previous method.

4. Conclusions

We have presented temperature and light dependent I-V characteristics of CVD synthesized few-layer MoS_2 with Ti/Au contacts. We have found a space-charge limited conduction consistent with the presence of an exponential distribution of trap states in the MoS_2 band-gap. We have studied the optical response of the MoS_2 flake applying a modified Hornbeck-Hayens model to study the photoconductive effect. Finally we have quantitatively evaluated the density of trap states in the range $10^{10} - 10^{11}\ cm^{-2}$.

References

[1] Di Bartolomeo A Genovese L Foller T Giubileo F Luongo G Croin L Liang S J Ang L K Schleberger M *Nanotechnology* **2017** 28 (21) 214002
[2] Di Bartolomeo A Grillo A Urban F Iemmo L Giubileo F Luongo G Amato G Croin L Sun L Liang S J Ang L K *Adv. Funct. Mater.* **2018** 28 (28) 1800657
[3] Urban F Passacantando M Giubileo F Iemmo L Di Bartolomeo A *Nanomaterials* **2018** 8 (3) 151
[4] Tao L *et al. ACS Appl. Mater. Interfaces* **2017** 9 12073
[5] Jariwala D *et al. ACS Nano* **2014** 8 (2) 1102-1120
[6] Mak K F Lee C Hone J Shan J Heinz T F *Phys. Rev. Lett.* **2010** 105 136805
[7] Ellis J K Lucero M J Scuseria G E *Appl. Phys. Lett.* **2011** 99 261908
[8] Di Bartolomeo A *et al. Appl. Phys. Lett.* **2016** 109 023510
[9] Giubileo F Di Bartolomeo A *Prog. Surf. Science* **2017** 92 (13) 143-175
[10] Di Bartolomeo A *et al. 2D Materials* **2018** 5 015014
[11] Giubileo F *et al. arXiv:*1808.01185 2018
[12] Lin M W *et al. J. Phys. D: Appl. Phys.* **2012** 45 345102
[13] Joung D Chunder A Zhai L Khondaker S I *Appl. Phys. Lett.* **2010** 97 093105
[14] Mark P Helfrich W *J. Appl. Phys.* **1962** 33 205
[15] de Boer R W I *et al. Phys. Status Solidi A* **2004** 201 1302
[16] Kumar V Jain S C Kapoor A K Poortmans J Mertens R *J. Appl. Phys.* **2003** 94 1283
[17] Takanashi Y Takahata K Muramoto *Y. IEEE Electron Device Lett.* **1998** 19 472-474
[18] Furchi M M Polyushkin D K Pospischil A Mueller T *Nano Lett.* **2014** 12 14(11) 6165-70
[19] Ghatak S Ghosh A *Appl. Phys. Lett.* **2013** 103 122103
[20] Di Bartolomeo A *et al. arXiv*:1808.02127v2
[21] Hornbeck J A Haynes J R *Physical Revivew* **1955** 97 311
[22] Wang R *et al. Phys. Rev. B* **2012** 86 045406

IOP Conf. Series: Journal of Physics: Conf. Series **1226** (2019) 012014 doi:10.1088/1742-6596/1226/1/012014

Monitoring the solid-state dewetting of densely packed arrays of Au nanoparticles

M Magnozzi[1], **M Ferrera**[1], **M Canepa**[1], **F Bisio**[2]

[1] OptMatLab, Dipartimento di Fisica, Università di Genova, via Dodecaneso 33, 16146 Genova, Italy

[2] CNR-SPIN, corso Perrone 24, 16152 Genova, Italy

E-mail: magnozzi@fisica.unige.it

Abstract. We report a real time, *in-situ* spectroscopic ellipsometry study of the temperature-induced solid-state dewetting of Au nanowires into nanoparticles. Very large spectral variations are observed at different temperatures. Analysis of the key features in the acquired spectra reveals two different regimes: up to 300 °C the variation in the optical response is dominated by solid-state dewetting, while above that temperature, smaller variations not compatible with such mechanism are visible. Therefore our ellipsometry measurements allow us to determine in real time at which temperature the solid-state dewetting ceases and the morphology of our sample becomes stable. We point out that this observation is possible thanks to the higher sensitiviy of ellipsometry with respect to reflectance/transmittance measurements.

1. Introduction

The drive toward the applications of metal nanostructures and nanoparticles calls for the availability of large-scale samples with uniform coverage and properties, that imply a high degree of spatial order. Among the bottom-up approaches commonly used to fabricate ordered nanostructures, the mechanisms which induce a self-organization of nanoparticles are subjected to a great deal of interest [1–5]. Self-organization means that nano-objects self-arrange in ordered structures (*e.g.* lattices, grids); for this reason, the fabrication process of self-assembled systems aims to control the driving forces causing the self-assembly, rather than the arrangement of single nano-objects [6].

Under proper conditions, solid-state dewetting can be exploited to induce the self-organization of nanoparticles. Solid-state dewetting causes agglomeration in thin films and nanostructures, at temperatures that are well below the melting temperature of bulk materials [7,8]. By exploiting nanopatterned substrates, it is possible to create ordered arrays of nanoparticles with tunable size and spacing [9]. Arrays of nanoparticles created by solid-state dewetting on our LiF nanopatterned substrates are characterized by high nanoparticles density ($\sim 10^{11}$ NPs/cm^2) and few-nm interparticle distance [10]. This kind of system has been used to investigate the Localized Surface Plasmon Resonance (LSPR) of noble metals and aluminum under diverse experimental conditions [11–14]. So far, in our research we focused on the outcome of the samples fabrication process (that is, the arrays of nanoparticles); however, useful information can be obtained by monitoring *in-situ* and in real time the evolution of the optical (plasmonic) response of our samples during their fabrication. By performing spectroscopic ellipsometry (SE) measurements during the temperature-induced solid-state dewetting in high vacuum, we noticed

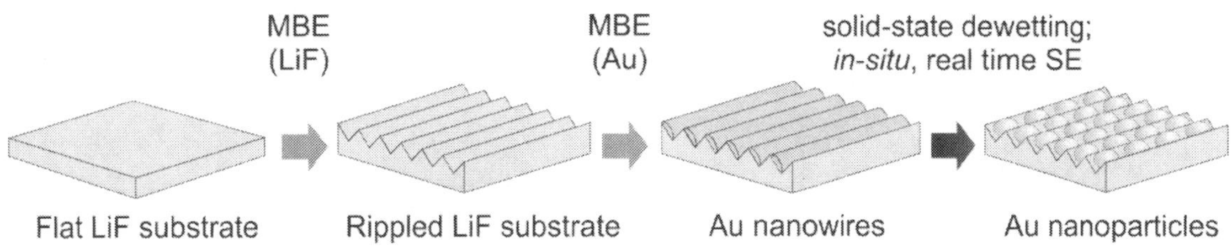

Figure 1: Sample preparation. From left to right: a crystal of LiF is rippled by depositing LiF *via* MBE; glancing angle MBE of Au creates disconnected Au nanowires; solid-state dewetting induces a transformation from nanowires to nanoparticles. This last process is monitored *in-situ* and in real time *via* spectroscopic ellipsometry.

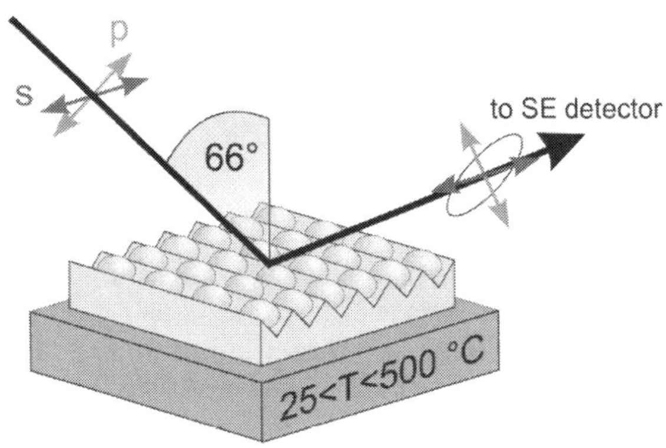

Figure 2: Sample orientation during the real time, *in-situ* SE measurements. Au NPs chains are parallel to the plane of incidence; angle of incidence is 66°. The sample is placed inside a high-vacuum chamber, and its temperature can be varied from 25 to 500 °C.

a huge variation in the key features of the samples' spectra; in particular, a large blueshift, and a progressive narrowing and deepening of the features linked to the LSPR were observed. Interestingly, this trend is monotonous only up to $\sim$ 300 °C; for higher temperatures, a small blueshift is still present, but the trend in the other two features is reversed. Therefore we identify two regimes: for lower temperatures, solid-state dewetting drives the variations in the optical response, while for higher temperatures, morphology remains largely stable and other effects become dominant.

2. Results and discussion

By following a well-established procedure, we fabricated a (1×1 cm^2) nanorippled LiF substrate (see Fig. 1) by depositing LiF *via* Molecular Beam Epitaxy (MBE) on a LiF (110) substrate in a dedicated high-vacuum chamber ("deposition chamber", base pressure $\sim$ 10^{-9} mbar) [11]; then in the same chamber, we deposited Au at glancing incidence to create Au nanowires as represented in Fig. 1. The sample was then extracted from the deposition chamber and immediately put inside a second high-vacuum chamber ("measurement chamber") to perform the real time, *in-situ* SE measurements. The measurement chamber has a base pressure of 10^{-7} mbar and allows the sample to be heated from room temperature to 500 °C; more details about the measurement chamber can be found in Ref. [15]. The ellipsometer (J.A. Woollam Co., M-2000) has a spectral range between 245 and 1450 nm; the acquisition of a full spectrum requires just a couple of seconds. The sample was heated with a temperature ramp (heating

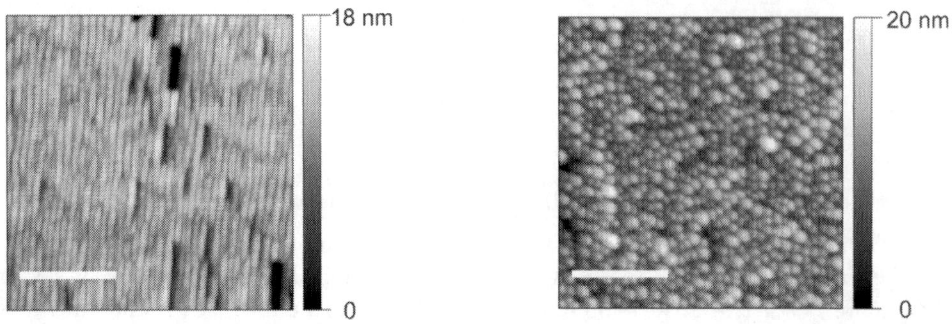

Figure 3: Left: representative AFM image of rippled LiF substrate. Right: Au nanoparticles as obtained after the solid-state dewetting of Au nanowires. White scale bar for both images is 400-nm long.

rate: 5 °C/min) and SE spectra were acquired from room temperature to 362 °C, at 12.5 °C interval. All SE measurements were performed at θ=66°, with the plane of incidence parallel to the direction of the ripples (see Fig. 2).

The morphology of our samples was investigated through Atomic Force Microscopy (AFM). AFM images were acquired in tapping mode using a Digital Instruments Dimension 3100 Atomic Force Microscope; image analysis was performed using the free software Gwyddion. Representative AFM images of our sample are shown in Fig. 3. On the left image is the rippled LiF surface; ripples periodicity is approximately 30 nm. The right image shows the sample surface after Au deposition and thermal annealing. Here we see nanoparticles aligned along the direction of the LiF ripples; mean interparticle distance is 33±2 nm along the ripples and 37±4 nm across them. As for the particles, their mean radius is 13±1 nm along the ripples and 11±1 nm across them. While AFM images are necessary to evaluate morphology, they are obtained only *before* or *after* the fabrication; this means that the intermediate variations induced on the sample by the thermal annealing are not accessible. In this regard real time, *in-situ* SE can provide a unique and valuable contribution.

The output of SE measurements is constituted by two angles called Ψ and Δ. In this work, we follow the definition

$$\rho = \tan \Psi \exp(i\Delta), \tag{1}$$

where $\rho = r_p/r_s$ and $r_{p,s}$ are the complex Fresnel reflection coefficients [16]. Thus $\Psi = \arctan(|r_p|/|r_s|)$ and $\Delta = \delta_p - \delta_s$, where $\delta_{p,s}$ are the phases of the p, s components of the reflected beam.

Real time, *in-situ* SE measurements are reported in Fig. 4 (panels A and B). Both Ψ and Δ spectra show macroscopic variations as the temperature rises. The most relevant feature in the Ψ spectrum at room temperature is the broad dip centered around 1000 nm, which is linked to the LSPR of the system [10]; for increasing temperatures, this dip steadily shifts towards shorter wavelengths and becomes deeper and narrower, meaning that the intensity of the LSPR increases. However, from 300 to 362 °C, this trend is reversed as the dip becomes significantly shallower and slightly broader (see Fig. 4A, inset). As far as Δ is concerned, at room temperature we observe a progressive rise in the Δ values when going from the visible to the IR; this becomes much sharper as temperature increases, until a phase inversion (i.e., Δ going to negative values) is observed.

An optical model is usually required to fully access the information contained in the Ψ and Δ spectra [16]. Modelling the system under investigation is a complex task that we discuss in detail elsewhere [17]; in this work, however, we present and discuss the useful information that can be extracted from our SE measurements without resorting to a model. Ψ and Δ spectra

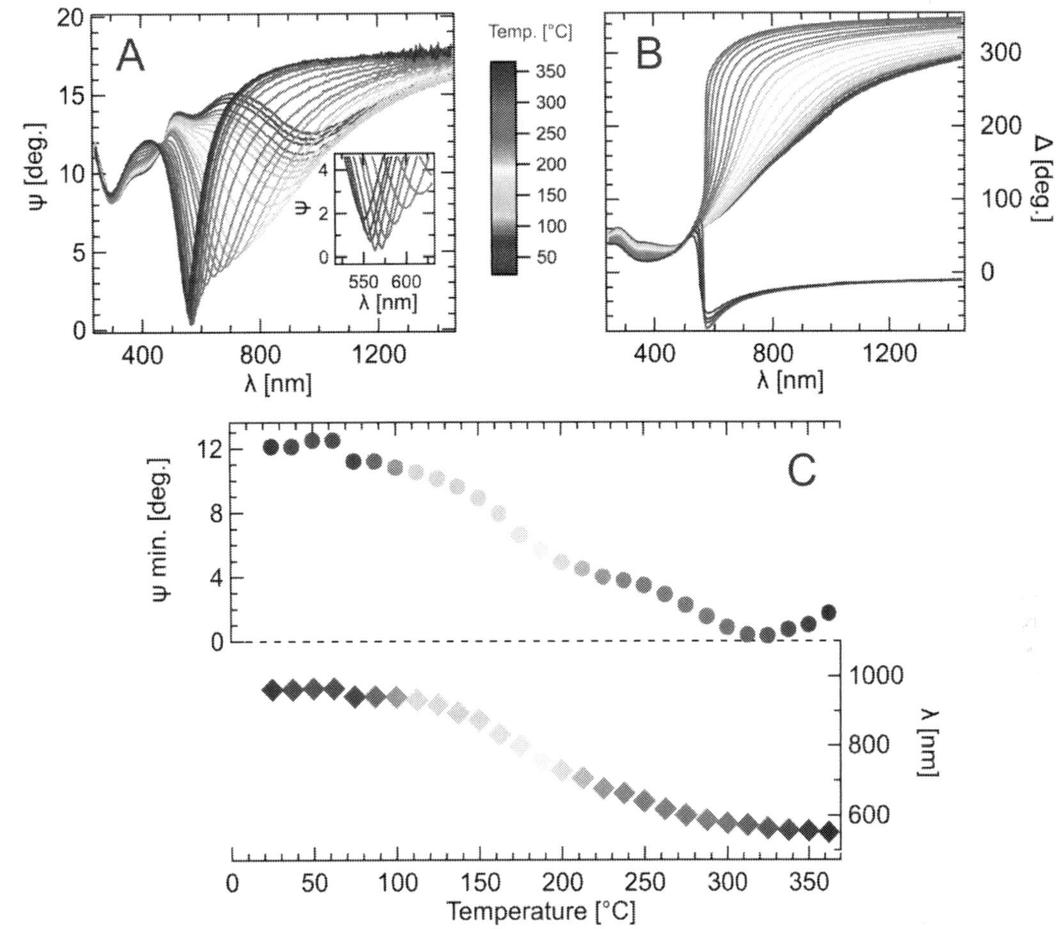

Figure 4: Panels A and B: Ψ and Δ spectra acquired *in-situ* and in real time during the dewetting of Au nanowires into nanoparticles. The color indicates the temperature, as shown in the color bar between the panels. Panel C: Left axis: intensity of the Ψ dip. Right axis: Spectral position of the Ψ dip.

are related to the complex Fresnel coefficients via Equation (1), therefore features linked to the LSPR can be seen in both spectra; however the dip in Ψ spectra offers a more intuitive way of understanding the morphological and physical changes induced in the sample by the thermal annealing. For this reason, in the following discussion we are going to focus on the Ψ dip.

In Fig. 4C we plot the spectral position and intensity of the Ψ dip. This plot reveals quantitatively the progressive blueshift of the LSPR, going from 960 nm at room temperature to 550 nm at 362 °C. Such a large spectral shift (410 nm) and all the other changes visible in Ψ and Δ spectra testify the major morphological change sustained by the system when Au nanowires dewet into densely-packed nanoparticles. We note that the blueshift is continuous as the temperature rises, and the largest variations occur approximately between 125 and 275 °C. However, the intensity of the Ψ dip does not follow a unique trend. We notice a very small rise at low temperatures, which is possibly due to desorption of ambient contaminants; then a continuous decrease from 11 degrees at 75 °C to almost 0 degrees at 300 °C; finally, from 300 to 362 °C, the trend is reversed as the minimum rises again. At the temperature where the dip in Ψ reaches its absolute minimum, Δ shows a sharp transition to negative values, assuming a wiggled shape.

A blueshift of the LSPR is observed in the longitudinal mode when plasmonic nanostructures reduce their aspect ratio [18]. In our case, the reduction in the aspect ratio occurs as a result of the solid-state dewetting of nanowires (high aspect ratio) into nanoparticles (low aspect ratio), as represented in Fig. 1 and 3. So we identify solid-state dewetting as the main cause of the large blueshift in the LSPR. However, solid-state dewetting cannot account for broadening and reduction of the Ψ dip (that is, a damping of the LSPR) at the highest temperatures; therefore, we have to invoke another mechanism that can explain such an inversion in the trend shown in Fig. 4. Surface melting, which is known to appear at temperatures much lower than the bulk melting [19, 20], increases the scattering rate of the free electrons, effectively damping the LSPR. The damping of the LSPR is opposite to the effect of solid-state dewetting; this means that at the temperatures where the damping is observed, the solid-state dewetting is no longer the main force driving the changes observed in the SE spectra. Solid-state dewetting may still be active from 300 to 362 °C, since there is a relatively small yet sizable (20 nm) blueshift of the dip in Ψ in that temperature range; however, subsequent heating/cooling cycles on the same sample showed that any further change in the SE spectra acquired at different temperatures is fully reversible, meaning that the morphology of our sample has found a stable configuration already at the end of the first thermal annealing.

3. Conclusions

We studied *via* real time, *in-situ* Spectroscopic Ellipsometry the evolution in the optical response of the sample during a thermal annealing, which induced a solid-state dewetting causing nanowires to transform into nanoparticles. By analyzing the spectral position, intensity and width of the Ψ dip related to the LSPR, we noticed two different trends in the evolution of the opical response of the sample. We identified solid-state dewetting as the main driver of change below 300 °C, while for higher temperatures, an opposite trend (damping of the LSPR) was observed in the SE spectra. In conclusion, from the analysis of the SE spectra we can determine the temperature at which the solid-state dewetting of Au is largely completed and our system becomes morphologically stable. In this way we can control in real time the fabrication of densely-packed Au NPs arrays. Lastly, we note that the observation of a clear difference between the two regimes discussed before is possible thanks to the fact that the LSPR feature in the SE spectra is very sharp and almost noise-free. Indeed, such an observation would have been virtually impossible in standard reflectance/transmittance measurements, because in that case the LSPR peak is much broader and noisy (see Ref. [10]).

References

[1] Grzelczak M, Vermant J, Furst E M and Liz-Marzán L M 2010 *ACS Nano* **4** 3591
[2] Kotov N A and Weiss P S 2014 *ACS Nano* **8** 3101
[3] Ozin G A, Hou K, Lotsch B V, Cademartiri L, Puzzo D P, Scotognella F, Ghadimi A and Thomson J 2009 *Mat. Today* **12** 12
[4] Karg M, Hellweg T and Mulvaney P 2011 *Adv. Funct. Mater.* **21** 4668
[5] Brasse Y, Müller M B, Karg M, Kuttner C, König T A F and Fery A 2018 *ACS Appl. Mat. Inter.* **10** 3133
[6] Ottewill R H, Schofield A B, Waters J A and Williams N S J 1997 *Colloid Polym. Sci.* **275** 274
[7] Leroy F, Borowik L, Cheynis F, Almadori Y, Curiotto S, Trautmann M, Barbé J C and Müller P 2016 *Surf. Sci. Rep.* **71** 391
[8] Thompson C V 2012 *Annu. Rev. Mater. Res.* **42** 399
[9] Anghinolfi L, Mattera L, Canepa M and Bisio F 2012 *Phys. Rev. B* **85** 235426
[10] Anghinolfi L 2012 *Self-organized arrays of gold nanoparticles. Morphology and plasmonic properties* (Berlin Heidelberg: Springer-Verlag)
[11] Anghinolfi L, Moroni R, Mattera L, Canepa M and Bisio F 2011 *J. Phys. Chem. C* **115** 14036

IOP Conf. Series: Journal of Physics: Conf. Series **1226** (2019) 012014 doi:10.1088/1742-6596/1226/1/012014

[12] Caminale M, Anghinolfi L, Magnano E, Bondino F, Canepa M, Mattera L and Bisio F 2013 *ACS Appl. Mater. Inter.* **5** 1955
[13] Maidecchi G, Gonella G, Proietti Zaccaria R, Moroni R, Anghinolfi L, Giglia A, Nannarone S, Mattera L, Dai H L, Canepa M and Bisio F 2013 *ACS Nano* **7** 5834
[14] Proietti Zaccaria R, Bisio F, Das G, Maidecchi G, Caminale M, Vu C D, De Angelis F, Di Fabrizio E, Toma A and Canepa M 2016 *ACS Appl. Mater. Inter.* **8** 8024
[15] Magnozzi M, Bisio F and Canepa M 2017 *Appl. Surf. Sci.* **421** 651
[16] Fujiwara H 2007 *Spectroscopic Ellipsometry. Principles and Applications* (Chichester: John Wiley & Sons)
[17] Magnozzi M, Ferrera M, Mattera L, Canepa M and Bisio F 2018 Submitted to *Nanoscale*
[18] Maradudin A A, Sambles J R and Barnes, W L (eds.) 2014 *Modern Plasmonics* (Amsterdam: Elsevier)
[19] Dash J G 1989 *Contemp. Phys.* **30** 89
[20] Plech A, Cerna R, Kotaidis V, Hudert F, Bartels A, Dekorsy T 2007 *Nano Lett.* **7** 1026

IOP Conf. Series: Journal of Physics: Conf. Series **1226** (2019) 012015 doi:10.1088/1742-6596/1226/1/012015

Topological phase diagram of coupled spinless p-wave superconductors

A Maiellaro, F Romeo, R Citro

Dipartimento di Fisica, Università degli Studi di Salerno, via Giovanni Paolo II, Fisicano (Sa) 132-I-84084, IT

Spin-CNR, via Giovanni Paolo II, Fisciano (Sa) 132-I-84084, IT

E-mail: amaiellaro@unisa.it

Abstract. We investigate the topological properties of a ladder given by two one dimensional p-wave superconductors coupled together site to site by transversal hopping t. For small t, we also derive an effective model whose simpler form is feasible to implement some braiding procedures proposed in the literature.

1. Introduction

Majorana fermions (Mfs), hosted in superconductors [1], may be used as qubit of fault-tolerant topological quantum computers. A crucial challenge towards the topological quantum computer is to implement quantum operation of nearly degenerate quantum states by a dynamical process of Majorana fermions; the principal one is the braiding dynamic in superconducting nanowires [2]. Several proposals have been proposed to perform this operation, for example changing the hopping t and pairing Δ on chains or between two chains of Kitaev can be used to perform the braiding [3]. Another possibility is crossing them [4] or pairing two wires in the topological phase with an extra wire in the trivial phase [5]. The braiding of Majoranas is another open problem of modern physics and several proposals have been made to this end, as a measured-based scheme for performing braiding operation and detecting the non-Abelian statistics of Mfs without moving or hybridizing them [6]. For example, one proposal is the measurement of conductance that is strictly connected to parity of the systems [5]. Another actual challenge, considered in some reviews, is the analysis of the non-adiabatic corrections in the braiding process of real systems [7].

Along this guide line, a crucial role is played by the study of the topological phases of the various systems that host such excitations and how these phases are modified by tuning the parameters which represent the direct experimental control of a possible future measurement of Majorana zero modes (MzMs). A contribution in this direction is given by the study of a ladder of two Kitaev chains [8].

Our aim in this work is to consider the most simple generalization of the Kitaev chain that exhibits Majorana zero modes. We consider two one dimensional p-wave superconductors that are coupled together site to site by an hopping term t. First of all, we study the topological phase of the system and we check the existence of Majorana zero modes and how they are modified by the transversal hopping, with respect to Mfs of the two separated chains.

We also propose an effective model for our system that leads itself to perform some braiding procedures proposed in literature [9].

2. Model

We consider a model in the continuum whose Hamiltonian describes two one dimensional p-wave superconductors coupled together by an hopping term of strength t:

$$H = \int \left\{ \left[\sum_{i=1,2} \widehat{\Psi}_i^\dagger(x) \left(\frac{p^2}{2m} - \mu \right) \widehat{\Psi}_i(x) + \Delta l \left(\widehat{\Psi}_i^\dagger(x) \partial_x \widehat{\Psi}_i^\dagger(x) + h.c. \right) \right] - t \left(\widehat{\Psi}_1^\dagger(x) \widehat{\Psi}_2(x) + \widehat{\Psi}_2^\dagger(x) \widehat{\Psi}_1(x) \right) \right\} dx \tag{1}$$

where $\widehat{\Psi}_i$ are the continuum field operators that create and destroy fermions on chain i, μ is the chemical potential, Δ is the pairing strength of the p-wave superconductor and l is the lattice constant in such a way Δ has energy dimension.
This Hamiltonian can be rewritten in the following matrix form:

$$H = \frac{1}{2} \int dx \begin{pmatrix} \widehat{\Psi}_1^\dagger(x) & \widehat{\Psi}_1(x) & \widehat{\Psi}_2^\dagger(x) & \widehat{\Psi}_2(x) \end{pmatrix} H_{BdG} \begin{pmatrix} \widehat{\Psi}_1(x) \\ \widehat{\Psi}_1^\dagger(x) \\ \widehat{\Psi}_2(x) \\ \widehat{\Psi}_2^\dagger(x) \end{pmatrix}$$

where H_{BdG} is the Bogoliubov de Gennes Hamiltonian, whose matrix representation is:

$$H_{BdG} = \begin{pmatrix} -\frac{\hbar^2}{2m}\partial_x^2 - \mu & \Delta l \partial_x & -t & 0 \\ -\Delta l \partial_x & \frac{\hbar^2}{2m}\partial_x^2 + \mu & 0 & t \\ -t & 0 & -\frac{\hbar^2}{2m}\partial_x^2 - \mu & \Delta l \partial_x \\ 0 & t & -\Delta l \partial_x & \frac{\hbar^2}{2m}\partial_x^2 + \mu \end{pmatrix} \tag{2}$$

here the Nambu representation has been used [10] and we have assumed that the two wires have the same chemical potential and pairing term (this second choice is made to guarantee an analytical resolution of equations).
The Bogoliubov-de Gennes Hamiltonian has also the particle-hole symmetry that exchanges electrons with holes and is represented by an antiunitary operator ρ that anti-commutates with H_{BdG}:

$$\rho H_{BdG} \rho^{-1} = -H_{BdG}$$

Because of this symmetry, the spectrum of H_{BdG} must be symmetric around zero energy; indeed, for every eigenvector ψ with energy E there will be a particle-hole symmetric eigenvector $\rho\psi$ with energy $-E$. In our basis this operator is given by:

$$\rho = (I \otimes \sigma_x) K \tag{3}$$

where K is the complex conjugation and σ_x is the Pauli matrix.

3. Topological phases: Analytical resolution

To study the topological phase of the system, we check when this model admits zero energy modes. So, we must solve the BdG equation for zero energy modes: $H_{BdG}\psi = 0$, where we call $\psi = \sum_{i=1,2} |i\rangle \otimes \begin{pmatrix} u_i & v_i \end{pmatrix}^T$ the topological spinor, u_1, v_1, u_2, v_2 are the BdG wavefunctions [11] and $|1\rangle = \begin{pmatrix} 1 & 0 \end{pmatrix}^T$, $|2\rangle = \begin{pmatrix} 0 & 1 \end{pmatrix}^T$. The substitution of Eq.(2) into the BdG equation for zero modes gives the following matrix equation:

$$\partial_x^2 \psi + \frac{2m\mu}{\hbar^2}\psi - \frac{2ml}{\hbar^2}\partial_x[\Delta(I \otimes \sigma_x)\psi] + \frac{2mt}{\hbar^2}(\sigma_x \otimes I)\psi = 0 \tag{4}$$

IOP Conf. Series: Journal of Physics: Conf. Series **1226** (2019) 012015 doi:10.1088/1742-6596/1226/1/012015

we first expand the spinor in the basis given by the eigenstates of $\sigma_x \otimes I$ $(\psi_1, \psi_2, \psi_3, \psi_4)$:

$$\psi = a\psi_1 + b\psi_2 + c\psi_3 + d\psi_4. \tag{5}$$

With this choice of basis we decouple the equations for the ladder into equations for two single wires with a shift in the chemical potential. In fact, the substitution of Eq.(5) into Eq.(4) gives:

$$\begin{cases} \partial_x^2 \begin{pmatrix} a \\ b \end{pmatrix} + k_1^2 \begin{pmatrix} a \\ b \end{pmatrix} - \frac{1}{\xi}\sigma_x \begin{pmatrix} a \\ b \end{pmatrix} = 0 \\ \partial_x^2 \begin{pmatrix} c \\ d \end{pmatrix} + k_2^2 \begin{pmatrix} c \\ d \end{pmatrix} - \frac{1}{\xi}\sigma_x \begin{pmatrix} c \\ d \end{pmatrix} = 0 \end{cases} \tag{6}$$

where $k_1^2 = \frac{2m(\mu+t)}{\hbar^2}$, $k_2^2 = \frac{2m(\mu-t)}{\hbar^2}$, $\frac{1}{\xi} = \frac{2ml\Delta}{\hbar^2}$ and ξ has the dimensions of a length.

We remark that Eq.(6) are exactly the equations of two single p-wave wires in which we have introduced an effective chemical potential: $\mu_{eff}^1 = \mu + t$, for the first of Eq.(6) and $\mu_{eff}^2 = \mu - t$, for the second of Eq.(6).

To solve Eq.(6) and compute the topological spinor we expand the vectors: $\begin{pmatrix} a & b \end{pmatrix}^T$, $\begin{pmatrix} c & d \end{pmatrix}^T$ in the basis of the eigenstates of σ_x in such a way we obtain two ordinary linear differential equations of the second order with constant coefficients. We get four independent solutions:

$$\phi_{1,1} = N_1[e^{\alpha_+^- x} - e^{(\alpha_+^- - \alpha_+^+)L} e^{\alpha_+^+ x}]\left[\begin{pmatrix} 1 \\ 1 \end{pmatrix} \otimes \begin{pmatrix} 1 \\ 1 \end{pmatrix}\right], \quad \phi_{1,2} = N_2[e^{\beta_+^- x} - e^{(\beta_+^- - \beta_+^+)L} e^{\beta_+^+ x}]\left[\begin{pmatrix} 1 \\ -1 \end{pmatrix} \otimes \begin{pmatrix} -1 \\ -1 \end{pmatrix}\right] \tag{7}$$

$$\phi_{2,1} = N_3[e^{\alpha_-^- x} - e^{\alpha_-^+ x}]\left[\begin{pmatrix} 1 \\ 1 \end{pmatrix} \otimes \begin{pmatrix} i \\ -i \end{pmatrix}\right], \quad \phi_{2,2} = N_4[e^{\beta_-^- x} - e^{\beta_-^+ x}]\left[\begin{pmatrix} 1 \\ -1 \end{pmatrix} \otimes \begin{pmatrix} i \\ -i \end{pmatrix}\right] \tag{8}$$

where N_1, N_2, N_3, N_4 are the normalization coefficients and their analytical expression is present in [12] and we don't report it here because it is of interest only for the numerical calculation, $\alpha_j^\pm = \frac{j \pm \sqrt{1-4(k_1\xi)^2}}{2\xi}$, $\beta_j^\pm = \frac{j \pm \sqrt{1-4(k_2\xi)^2}}{2\xi}$, $j = \pm 1$. These solutions are obtained imposing the boundary conditions, the particle-hole symmetry and the normalization:

$$\phi_{1,1}(L) = \phi_{1,2}(L) = 0, \quad \phi_{2,1}(0) = \phi_{2,2}(0) = 0, \quad \rho\phi_{i,j} = \phi_{i,j}; \quad \int_0^L |\phi_{i,j}|^2 \, dx = 1$$

where $i, j = 1, 2$.

The modes in Eq. (7) are localized at the right-hand side of the ladder and those in Eq. (8) are localized at the left-hand side of the ladder. This distinction between four independent modes is made because, as we have already observed, α and β are the solutions of two indipendet wires with a shift in the chemical potential and because we have imposed that they are real in the topological phase requirement[1].

A simple remark is needed: these four modes are not the modes of the two single wires of the original Hamiltonian but are a linear combination of them.

4. Numerical results

We study the topological phase diagram of the model by tuning the transversal hopping t.

As is depicted in Fig.(1) and Fig.(2), when t increases we have a phase transition between two topological phases, one with four modes and one with two modes. This is because the stronger the transversal hopping, the greater the coupling between the two wires of the ladder and so it appears as a single wire with an internal structure and therefore exhibits only two topological modes.

[1] This last requirement implies that $k_1\xi < \frac{1}{2}$, $k_2\xi < \frac{1}{2}$.

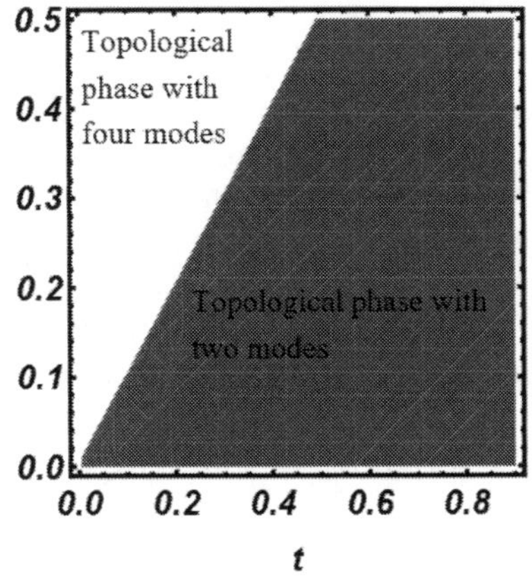

Figure 1. Topological phase diagram in the $\mu - t$ plane in unit of Δ with the fixed parameter $\frac{L}{\xi} = 75$. In the region of interest we can distinguish between two regions: a topological phase with four modes and a topological region with two modes.

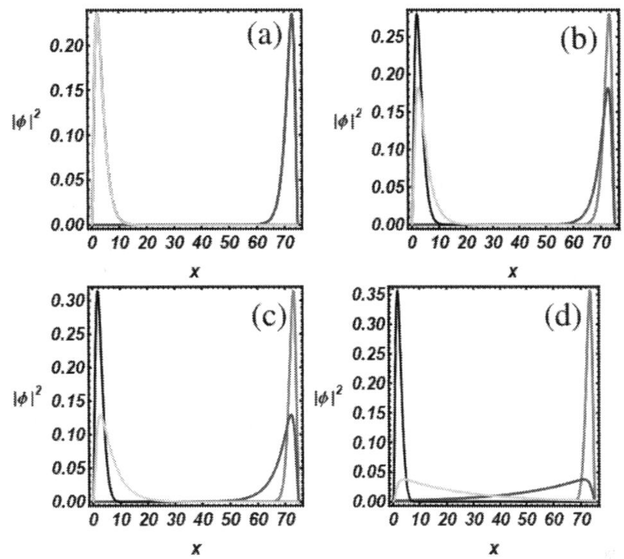

Figure 2. The square modulus of the four modes at increasing t. This plots are realized with: $\xi = 1$, $\frac{L}{\xi} = 75$ and $k_F\xi = \frac{\sqrt{2m\mu}}{\hbar} = 0.45$ and for $\frac{t}{\mu} = 0, \ 0.3, \ 0.55, \ 0.9$ respectively fot the panels (a), (b), (c), (d). In the panel (a) the four modes are overlapped, in the other panels two of the four modes are delocalizing. In the panel (d) two modes are recombining themselves and the model exhibits only two topological modes.

5. Topological effective Hamiltonian

For small t, we can consider the projection of Hamiltonian of Eq.(2) onto the Hilbert subspace identified by the four localized eigenstates of two separated wires. This new Hamiltonian will represent an effective Hamiltonian for the system; in fact, in the limit in which the system remains in the zero-degenerate variety the mapping $H \to H_{eff}$ is a change of basis. The utility of this Hamiltonian is, as we will see, that in its regime of validity it has a simpler expression than Eq.(2) and exhibits, for a long ladder, four "Majorana zero energy modes" (MzMs) and then it can be used to perform braiding procedures [9].
Introducing the projection operator onto our new basis $\Pi = \sum_{i=1}^{4} |i\rangle\langle i|$, we get:

$$H_{eff} = \Pi H_{BdG} \Pi = \sum_{i,j} H_{i,j} |i\rangle\langle j| \tag{9}$$

for $i, j = 1, .., 4$ and where $H_{i,j} = \langle i|H_{BdG}|i\rangle$ are the Hamiltonian matrix elements in the new basis. The states $|1\rangle, \ |2\rangle, \ |3\rangle, \ |4\rangle$ are the MzMs of two separated wires and are well-defined in literature [12]. Using the expression of the four modes of the two separated wires we numerically calculate the matrix elements of Eq.(9) and we get (from now on we express the parameters of

Hamiltonian in unit of Δ and l.):

$$H_{eff} = \begin{pmatrix} 0 & -ib & 0 & -ict \\ ib & 0 & ict & 0 \\ 0 & -ict & 0 & -ib \\ ict & 0 & ib & 0 \end{pmatrix} \tag{10}$$

where we have explicited the dependence on the transversal hopping t and included the other dependencies inside the coefficients $b \equiv b(L, \ \xi, \ \mu)$ and $c \equiv c(L, \ \xi, \ \mu)$. The analytical expressions of c and d are present in [12]. Numerical diagonalization of Eq.(10) gives us the eigenvalues and eigenvectors.

As is depicted in Fig.(3), we have a pair of values of μ and t for which two modes have exactly

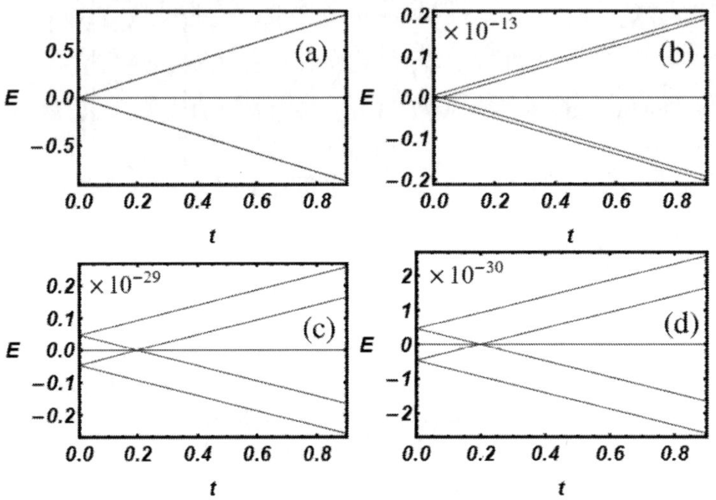

Figure 3. Plots of eigenvalues of H_{eff} in function of the transversal hopping t for different values of chemical potential μ. The plots are realized in unit of Δ, for fixed $\frac{\xi}{l} = 10$, $\frac{L}{l} = 750$ and for: $\frac{\mu}{\Delta} = 0, \ \ 0.05, \ \ 0.2, \ \ 0.45$ respectively for the panels (a), (b), (c), (d).

zero energy, in particular for $\mu = 0$, the effect of transverse hopping is that of removing two modes from zero while when μ increases there is an exact value of t for which two modes go to zero.

Indeed, in Fig.(3) we can see that all the four modes have a very small energy, in such a way we can approximately interpret all modes are Majorana-like modes. The regime of validity of H_{eff} is along the vertical cuts for small t in the phase diagram of Fig.(1). In this limit, we can consider the linear combination of the eigenstates of H_{eff} in Eq.(11) that are quasi-MzMs. Fig.(4) exhibits the plots of the square modulus of this quasi-MzMs. The particle-hole operator in this new basis: $\rho = IK$ confirms the Majorana character of this modes.

$$\begin{aligned} |\psi_1\rangle + |\psi_2\rangle &= \frac{1}{\sqrt{2}}\left[\begin{pmatrix} 1 \\ 1 \end{pmatrix} \otimes \begin{pmatrix} 0 \\ 1 \end{pmatrix}\right], \ \ |\psi_1\rangle - |\psi_2\rangle = \frac{1}{\sqrt{2}}\left[\begin{pmatrix} 1 \\ 1 \end{pmatrix} \otimes \begin{pmatrix} -i \\ 0 \end{pmatrix}\right], \\ |\psi_3\rangle + |\psi_4\rangle &= \frac{1}{\sqrt{2}}\left[\begin{pmatrix} 1 \\ -1 \end{pmatrix} \otimes \begin{pmatrix} 0 \\ -1 \end{pmatrix}\right], \ \ |\psi_3\rangle - |\psi_4\rangle = \frac{1}{\sqrt{2}}\left[\begin{pmatrix} 1 \\ -1 \end{pmatrix} \otimes \begin{pmatrix} -i \\ 0 \end{pmatrix}\right] \end{aligned} \tag{11}$$

where $|\psi_1\rangle, \ \ |\psi_2\rangle, \ \ |\psi_3\rangle, \ \ |\psi_4\rangle$ are the eigenvectors of Eq.(10).

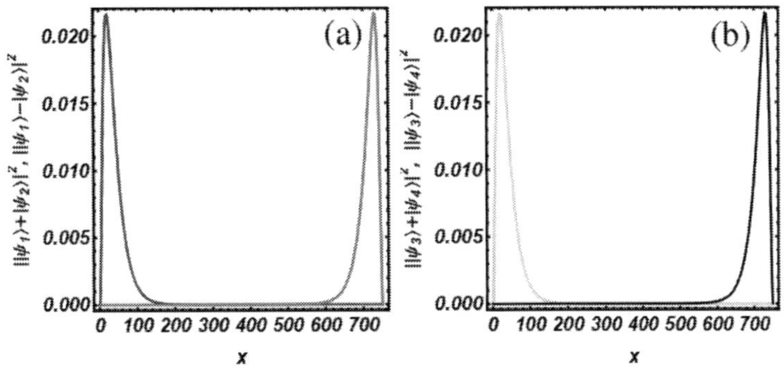

Figure 4. Square modulus of the linear combination of eigenstates of H_{eff}. We have that: $||\psi_1\rangle + |\psi_2\rangle|^2$ is the blue mode, $||\psi_1\rangle - |\psi_2\rangle|^2$ is the red mode, $||\psi_3\rangle + |\psi_4\rangle|^2$ is the green mode and $||\psi_3\rangle - |\psi_4\rangle|^2$ is the black mode. The plots are realized in units of Δ, for fixed $\frac{\xi}{l} = 10$, $\frac{L}{l} = 750$.

6. Conclusions

In this work, we have presented a generalization of the well known topological model of a one dimensional p-wave superconductor to the case of two p-wave superconductors coupled together site to site via transversal hopping. We we have found that when t increases the system exhibits two topological phases one with four zero energy modes and one with two zero energy modes as the effect of a stronger coupling. This is a check of the robustness of topological phase against possible perturbations lead by the variation of parameters. We have also derived an effective model that exhibits four Majorana zero energy modes. In the limit of small t, this effective model represents a good candidate to perform some braiding procedures proposed in literature [9].

7. Acknowledgements

R.C. acknowledges the project Quantox (Grant Agreement N. 731473) of Horizon 2020 Programme.

8. References

[1] Oppen F *et al.* *Topological superconducting phases in one dimension*, Oxford press: Lecture notes, 2014
[2] Pachos J K, *Introduction to topological quantum computation*, Cambridge University University Press, Cambridge, 2012
[3] Laflemme C *et al.* Phys Rev A **89**, 022319
[4] Karzig T *et al.* Phys Rev B **91** 201102
[5] Amorim C S *et al.* Phys Rev B **91** 174305
[6] Vijay S *et al.* Phys Rev B **94** 235446
[7] Cheng M *et al.* Phys Rev X **3**, 041017
[8] Maiellaro A *et al.* Eur Phys J Spec Top, 227: 1397
[9] Alicea J *et al.* Nature Physic **7**, 412-417
[10] Aguado R, La Rivista del Nuovo Cimento **40**, 11 523-593
[11] Blonder G E *et al.* Phys Rev B, **25** 4515 (1982).
[12] Maiellaro A, *Probing and Manipulation of Majorana fermions in coupled Kitaev chains*, master thesis, 2017.

IOP Conf. Series: Journal of Physics: Conf. Series **1226** (2019) 012016 doi:10.1088/1742-6596/1226/1/012016

A computational investigation on the electronic and optical properties of Coronene and its Boron-Nitride and perfluorinated counterparts

P Mocci, R Cardia and G Cappellini

Department of Physics, University of Cagliari, S.P. Monserrato-Sestu Km 0,700 I-09042 Monserrato (CA), ITALY

E-mail: paola.mocci@dsf.unica.it

Abstract. We present a computational study on the electronic and optical properties of a representative C-made and Boron-Nitride-made (BN) planar molecule of interest for potential applications in the solid state domain. In particular, we analyzed the case of Coronene ($C_{24}H_{12}$) in its BN and perfluorinated analogues. We performed all electrons Density Functional Theory (DFT) and Time Dependent-DFT (TD-DFT) calculations using a localized Gaussian basis-set in combination with a hybrid exchange-correlation functional. For all the systems we have calculated different electronic properties and the optical absorption spectra. A discussion on the possible implications of the general trends, observed for the BN-made clusters properties as compared to their C-based parents, will be given.

1. Introduction

During the recent past, in the optoelectronics branch, a high interest and a deep attention has been covered by small molecules, known as Polycyclic Aromatic Hydrocarbons (PAHs), playing a meaning role for both the unrestrainable growth of the modern application areas and the theoretical research field. These systems are commonly employed as active elements in several optoelectronic devices, from light-emitting diodes and photovoltaic cells to liquid crystals and organic thin-film field-effect transistors [1]. Indeed, they present various advantages, such as: the possibility to be easily purified through evaporation and solution processings, accompained with their tunable electronic and optical features. In particular, among these compounds, the position occupied by the Circumacenes (Coronene, Ovalene, Circumanthracene, Circumtetracene, and Circumpentacene are the first five members of the family) has recently shown up, thanks to their attractive properties [2, 3]. Aside from their characteristic planar and symmetric geometry, these molecules potentially play a significant role as finite portions (with a nanometric size) of an infinite system widely studied and analyzed: the Graphene [4].
Furthermore, related to the surge of Graphene and Carbon-based material research, a vivid interest in the investigation of other 2D nanomaterials has soon manifested, for example on the 2D Boron-Nitride (BN) [5]. Bulk-BN, in fact, occupies a unique position, sharing many of its properties, structures, processing and applications with carbon. In particular, for cubic-BN ($c-$BN), its extreme hardness, wide energy band gap, low dielectric constant, and high

thermal conductivity are also very near to those of diamond [6]. 2D BN-based nanostructures are therefore considered as promising candidates for different technologic devices and applications: ranging from vacuum technology, microelectronic devices, wear-resistant lubrificants to x-ray lithography masks, deep UV LED, graphene-engineering [6] and biological and medical fields applications [7]. In this last context, the synthesized BN-nanotubes (BNNTs) were investigated for their potential use in biomedical applications as gene and drug transporter or tissue scaffolds after the evaluation of their non-toxicity and their biocompatibility [8]. The BN compound, formed by the union of Boron and Nitrogen atoms in the same percentage, shows up as a thermally and chemically resistant material. It presents various crystalline forms, isoelectronic with respect to the elementaty Carbon-made structure, among which:

- the hexagonal form (*h*-BN), that is the counterpart of the graphite (the most stable and soft form among BN polymorphs, usually employed as lubrificant, additive to cosmetic products, ceramics..);
- the cubic form (*c*-BN) which is the diamond analogue (softer than diamond but with a superior thermal and chemical stability, usually used as abrasive);

On the other hand, the differences between the BN-made compounds and the C-based ones are mostly due to the different nature of their chemical bond: for example, while the *h*-BN is an insulator, the graphite is a conductive semimetal [9].

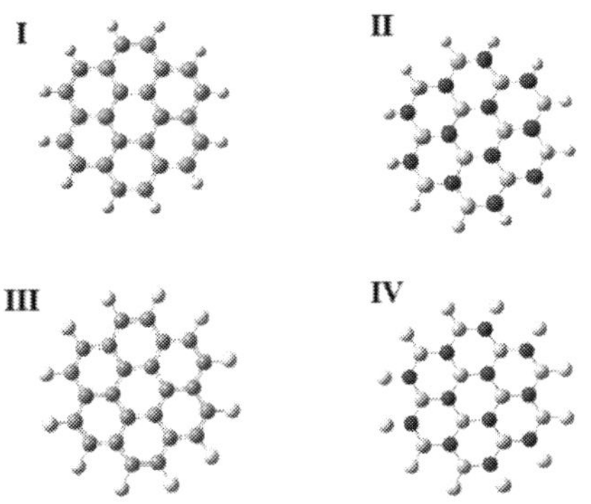

Figure 1. Coronene molecule (I) in its BN (II), perfluorinanted (III) and perfluorineted-BN-made (IV) counterparts. C atoms are indicated in grey, H in white, F in light-blue, N in blue and B in pink.

In the present work we expose a computational comparative analysis on the electronic and optical properties of some representative BN planar molecules of interest for potential applications in condensed matter physics domain [10, 11, 12]. In particular, we have selected the Coronene molecule (the first member among the Circumacenes family), studying its BN-made and perfluorinated counterparts [1]. We have performed Density Functional Theory (DFT)[13] and the Time Dependent-DFT (TD-DFT) calculations [14] to obtain the ground-state electronic properties and the excited-ones, respectively. We will discuss our results with existing previous theoretical and experimental data from the literature, focusing on the possible implications of the BN clusters properties in comparison with their C-made counterpart.

[1] The perfluorination corresponds to the total substitution of H atoms with F atoms in the peripheral molecular areas. A similar analysis has been made by Ref. [15].

IOP Conf. Series: Journal of Physics: Conf. Series **1226** (2019) 012016 doi:10.1088/1742-6596/1226/1/012016

2. Computational methods

According to previous works [1, 2, 16], geometry optimizations have been performed using a localized gaussian basis-set, $6-31+G^{*}$, combined with a hybrid exchange-correlation functional B3LYP [17, 18, 19], within an all electrons computational code, *Gaussian*16 [20]. The selected basis-set is composed by a valence double-ζ set augmented with d polarization functions for each atom (the + sign indicate the presence of the diffuse functions). The choice for the B3LYP as exchange-correlation potential, has been justyfied, because, with respect to other possibilities (e.g. the PBE - Perdew-Burke-Ernzerhof [21]), it has been demonstred to reproduce better results for different clusters of PAHs molecules, for both the ground-state and the excited one [1, 23, 25, 27, 28].

We have performed the DFT [13] and Time dependet-DFT (TD-DFT) [14] calculations implemented in the *Gaussian*16 computational package [20]. In particular, through the DFT scheme we have obtained the ground-state and the electronic properties (e.g electron affinities, ionization energies and quasi-particle (QP) gaps) for all the systems under study. While, the time-dependet counterpart of this method has allowed us to calculate the optical properties (optical onsets, exciton binding energies) and obtain the absoption spectra in the visible up to the middle ultraviolet (MUV) and far ultraviolet (FUV). In particular, we used the Casida computational scheme which computes the poles of the polarizability in the frequency domain (these poles correspond to the vertical excitation energies, while their strenghts represent the oscillator strenghts [22]).

Finally, we used the ΔSCF scheme [26, 29]. Within this method the vertical electron affinities (EA_V) and ionization energies (IE_V) can be calculated as differences between the ground-state total energy of the neutral system, E_n^0, and the energies of the charged species (the anion E_a^0 and the cation E_c^0, respectively), in correspondence of the neutral geometry.

This permits the evaluation of the quasi-particle (QP) gap (known also as "fundamental gap"), which in the ΔSCF scheme is defined as follows:

$$E_{gap} = IE_V - EA_V = (E_c^0 - E_n^0) - (E_n^0 - E_a^0) \quad (1)$$

In addition, knowing the first optically active transition from the absorption spectrum, E_{opt}, we could give an estimate of the exciton-binding energy through the difference: $E_{bind} = E_{gap} - E_{opt}$.

3. Electronic properties

In the following discussion, for brevity, the analyzed molecules will be indicated respectively by the acronyms in brackets: Coronene (C), BN-Coronene (BNC), perfluorinated-Coronene (p-C) and perfluorinated-BNCoronene (p-BNC). Tab. 1 presents the computed data of the electronic properties for all the molecules under study. For each observable the percentage variation, calculated with respect to the original parent (C), has been reported in brackets (for further details see the footnote [2]). Some of the data (vertical electron affinity (EA_V), vertical ionization energy (IE_V) and fundamental gap (E_{gap}) are shown in Fig.2 (A) through a diagrammatic representation.

In what follows, we have decided to focus in detail only in a comparative description on the electronic properties behavior, analyzing, in particular, what happens in the passage from C to BNC and from BNC to p-BNC. Considering the vertical electron affinity, going from C to BNC, a decrease nearly ~-200% takes place. On the other hand, the perfluorination effect in BNC determines a large increase (of the order of +200%) of the same observable. On the contrary, for the vertical ionization energy, in the passage from C to its BN-counterpart, we have registered a rise of about 20%. For this observable, the presence of Fluorine (F) atoms

[2] We report our calculated data with two significant digits since the precision of the experimental values available in the literature are of the order of 10 meV.

Table 1. Vertical electron affinity (EA_V), vertical ionization energy (IE_V) and fundamental gap (E_{gap}) for C, BNC, p-C and p-BNC. The (%) variation of each observable is computed with respect to the original C-made molecule (C). All the values are given in eV.

Molecule	EA_V	IE_V	E_{gap}
C	0.39	7.07	6.68
BNC	-0.26 (-167%)	8.45 (+19.5%)	8.71 (+30.4%)
p-C	1.77 (+354%)	8.25 (+16.7%)	6.49 (-2.8%)
p-BNC	0.35 (-10.3%)	9.68 (+36.9%)	9.33 (+39.7%)

produced a rise of similar entity in p-BNC (of about 15%) as compared to the corresponding original parent (BNC). For what concerns the fundamental gap, we observed an increase of ~ +30% from C to BNC. In this case, for p-BNC the fundamental gap increases of ~+7% with respect to its original analogue (BNC). Globally, the transition from the C-made molecule to the BN-made one provokes a lowering in the vertical affinity, an increase of the ionization energy and a consequent rise of the QP gap.The perfluorination effect going to BNC to p-BNC provokes a general raise of all the above mentioned observables.

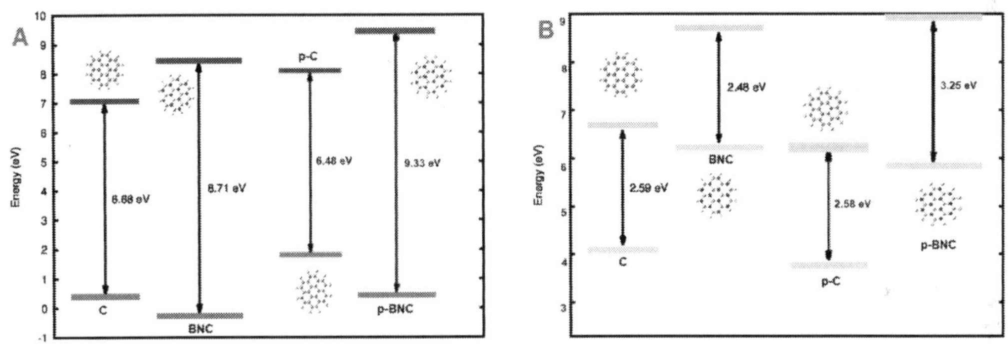

Figure 2. (A)Vertical electron affinities (red lines), vertical ionization energies (blue lines) and fundamental gaps (black arrows) for C, BNC, p-C and p-BNC, from left to right, respectively.
(B) Fundamental gaps (green lines), optical onsets (light-blue lines) and exciton binding energies (black arrows) for C, BNC, p-C and p-BNC, from left to right, respectively.

4. Optical properties

Tab. 2 reports the results for the optical properties, as obtained after TD-DFT calculations, while Fig. 2 (B) shows the trend of some of the previous observables through a diagramatic representation. Also in this case, we focus in detail on the comparison between C and BNC and BNC and p-BNC (the others molecular comparisons could be found referring to Tab. 2).

From the computed data we noted that both the main peak and the optical onset of BNC are blueshifted of ~80% and ~ 50%, respectively, as compared to those of C. For what concerns the perfluorination effect, the presence of F-atoms implicates a redshift of the optical onset and the dominant peak in the p-BNC with respect to BNC and respectively of the order of 2% and 14%. Globally, the absorption edge takes place in the UV for all the examined molecules, accompained with a blueshift and a remodeling of the absorption structure for the BN-systems with respect to that of their C-made original parent.

Table 2. Energies relative to the main peak position (M.P.P.), optical onset (E_{opt}) and exciton-binding energy (E_{bind}) for C, BNC, p-C and p-BNC. The% variation of each observable is calculated with respect to the original C-made parent molecule. All values are given in eV.

Molecule	M.P.P.	E_{opt}	E_{bind}
C	4.09	4.09	2.59
BNC	7.44 (+82%)	6.22 (+52%)	2.49 (-4%)
p-C	6.44 (+57%)	3.91 (-4%)	2.59 (-)
p-BNC	6.38 (+56%)	6.07 (+48%)	3.26 (+26%)

For what concerns the exciton-binding energy this observable remains almost constant from a molecule to an other: in particular, a negligible reduction ($\sim$ 4%) from C to BNC takes place, while a rise of $\sim$ 26% is determined as a consequence of the perfluorination in BNC.

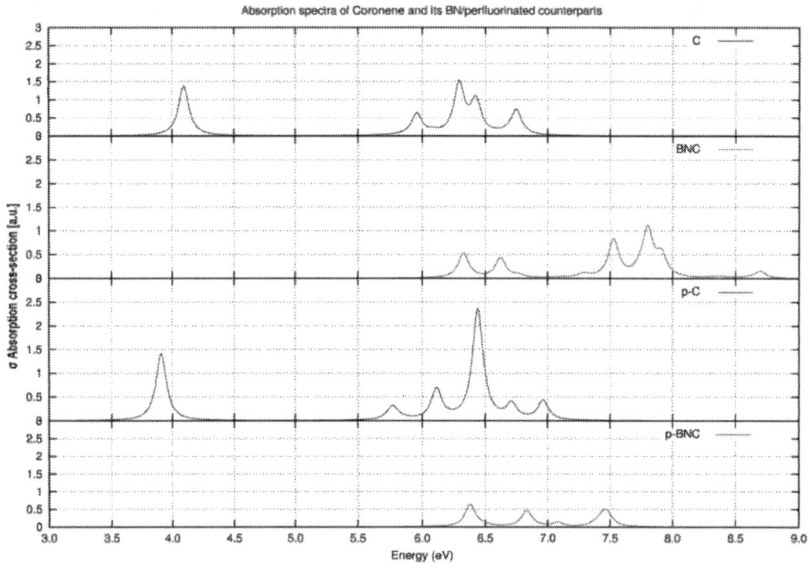

Figure 3. Absorption spectra (absorption cross-section [a.u.] vs energy [eV]) for C, BNC, p-C and p- BNC, from the top to the bottom, respectively.

Moreover, from Fig. 3 the optical onsets of the BNC and p-BNC are blueshifted with respect to their C-made original parents: the absorption is mainly translated towards the UV region. The typical absorption structure of the C-based material, with a peak in the region close to 4 eV, disappears in the BN-made molecules spectra, with a redistribution of the intensity at higher energies. The presence of F-atoms in the BN-system, produces a redshift for both the main peak (from 7.44 to 6.38 eV) and for the optical onset (6.22 from to 6.07 eV), as previously discussed.

Conclusions

We have presented a comparative study of Coronene and its BN and perfluorinated forms in the context of DFT and TD-DFT techniques. Dealing with the electronic properties, we have found that a lowering of the vertical electron affinities and a rise of the ionization energies and

QP gaps take place passing from the C-made to the BN-based systems.
For what concerns the optical properties, we observed that the absorption is localized in the UV for all the molecules under study, with a blueshift of the optical onsets and the dominant peaks for the BN-made molecules with respect to the original C-made parents. Moreover, the perfluorination effect has similar entity either for the Carbon and the Boron-Nitride compounds. This quantitative comparison on the electronic and optical properties of these compounds, could open future ways on a theoretical level, in order to take full advantages in many application fields as emerging novel materials. The high chemical inertia, thermal stability, the tunable electronic and optical properties, no-toxicity, make these BN-nanostructures as promising candidates for different technologic devices, complementary but also innovative with respect to the successfull Carbon-based counterpart [7]. In particular, thanks to their UV-absorption feature here analysed, these BN nanomaterials keep a promising role in the technology linked to UV optoelectronics.

Acknowledgments

P M gratefully acknowledges Sardinia Regional Government for the financial support of her PhD scolarship (P.O.R.Sardegna F.S.E. Operational Programme of the Autonomous Region of Sardegna, European Social Fund 2007-2013 Axis IV Human Resources, Objective I.3, Line of Activity I.3.1). G C and R C acknowledge partial financial support from IDEA-AISBL Bruxelles. Moreover, G C acknowledges partial financial support from "Progetto biennale d'Ateneo UniCa/FdS/RAS (Legge Regionale 7 agosto 2007, n. 7– annualitá; 2016)-Multiphysics theoretical approach to thermoelectricity".

References

[1] Malloci G, Cappellini G, Mulas G and Mattoni A 2011 *Chem. Phys.* **384** 19-27
[2] Cardia R, Malloci G, Cappellini G and Serra G 2016 Proc. of SPIE 9895 302-308
[3] Perez-Jimenez A J and Sancho-Garcia J-C 2009 IOP *Publishing Ltd Nanotechnology* **20** 47
[4] Geim A and Novoselov K 2010 *Random walk to Graphene* - The Nobel Prize Lecture
[5] Filippetti A, Fiorentini V, Cappellini G and Bosin A 1998 *Physica Stuatus Solidi. A, Appl. Res.* **170** 265-269
[6] Cappellini G, Satta G, Palummo M and Onida G 2001 *Phys. Rev.* **64** 035104 1-11
[7] Lin Y and Connell J W 2012 *Nanoscale* **4** 6908
[8] Kalay S *et al* 2015 *Beilstein J. Nanotechnol.* **6** 84102
[9] Bassani F and Grassano U 2000 *Fisica dello stato solido* Bollati Boringhieri
[10] Sandoval E D *et al* 2016 *Phys. Rev.* B **94**
[11] Bernardi M, Palummo M, Grossman J C 2012 *Phys. Rev. Letters* **108** 226805
[12] Nemnes G A *et al* 2016 *J. Phys.: Conf. Ser.* **738** 012027
[13] Kohn W 1999 *Rev. Mod. Phys.* **71** 1253
[14] Runge E and Gross E K U 1984 *Phys. Rev. Letters.* **52** 997
[15] Lai L *et al* 2006 *J. Phys Chem. B* **110** 14092-14097
[16] Mocci P, Cardia R and Cappellini G 2018 *The Journal of Physics- Conference Series* **956** 012020
[17] Becke A D 1993 *J. Chem. Phys.* **98** 5648
[18] Lee C, Yang W and Parr R G 1988 *Phys. Rev. B* **37** 785
[19] Stephens P J, Devlin P J, Chabalowski C F, and Frisch M J 1994 *J. Phys. Chem.* **98** 11623
[20] M J Frisch *et al* 2016 Gaussian, Inc., Wallingford CT.
[21] Ernzerhof M *et al* 1999 *Journal of Chem. Phys.* **110** 11
[22] Casida M E 1995 *Time-Dependent Density Functional Response Theory for Molecules* World Scientific**1**
[23] Malloci G, Cappellini G, Mulas G, Mattoni A 2011 *Chem. Phys.* **384** 19-27
[24] Cardia R, Malloci G, Mattoni A, Cappellini G 2014 *J. Phys Chem. A* **118** 5170-5177
[25] Cardia R, Malloci G, Rignanese GM, Blase X, Molteni E, Cappellini G 2016 *Phys.Rev. B* **3** (23) 235132
[26] Malloci G, Cappellini G, Mulas G and Satta G 2004 *Phys. Rev. B* **70** 205429-1
[27] Cappellini G, Malloci G and Mulas G 2009 *Superlattices and Microstruc.* **46** 14-18
[28] Kumar A *et al* 2018 *Cellulose* **25** 2119-2203
[29] Jones R O and Gunnarsson O 1989 *Rev. Mod. Phys.* **61** 689-746

IOP Conf. Series: Journal of Physics: Conf. Series **1226** (2019) 012017 doi:10.1088/1742-6596/1226/1/012017

The build up of a test station facility for the SIS100 superconducting modules

U Gambardella[1], G Iannone[1] and A Saggese[1,2]

[1] INFN-Napoli, at Physics Department, University of Salerno, Fisciano 84084, Italy
[2] Physics Department, University of Salerno, Fisciano 84084, Italy

E-mail: saggese@unisa.it

Abstract. More than eighty quadrupole modules, especially designed and fabricated at JNIR in Dubna, will be used at the FAIR facility in Darmstadt. Prior the installation in the new accelerator SIS 100 tunnel, the modules need to be tested by an independent laboratory. The required test have to be the same for all the modules and the answer will be the benchmark for the final decision on accepting or rejecting the module. With this target in mind, a system with as much as possible automatisms will be realized. Three kinds of tests will be performed: mechanical, thermal and electrical. The mechanical test will concern on the search of vacuum leaks, if present, and on the measurement of the limit pressure reached in the vessel. The thermal test will have to check that the various thermometer, scattered inside the modules, will be correctly working and that the heaters are able to deliver the generated heat. The electrical tests will verify the insulation to ground of both the devices and the superconducting coils and bus-bars. All tests will be done in a thermal sequence: at room temperature, at the cryogenic working temperature of 4.5 K, and finally at room temperature again.

1. Introduction

The GSI (Helmholtzzentrum für Schwerionenforschung GmbH) is a company associated with the German Federal Government based near Darmstadt in the region of Hesse, in Germany. This laboratory performs basic and applied research in physics and related natural science disciplines using the heavy ion accelerator facility. Within the GSI, the new FAIR (Facility for Antiproton and Ion Research) accelerator complex is under construction. The FAIR facility is a new unique international accelerator complex for the research with the antiprotons and heavy ions. From the construction point of view, the realization of this accelerator is quite different than the other large particles accelerators, *e.g.* the CERN (European Organization for Nuclear Research) LHC (Large Hadron Collider) accelerator ring. In fact, this accelerator complex is based on magnets ramped up to 4 T/s, fast respect to the corresponding LHC magnets that are ramped at only 0.01 T/s.

In the framework of the collaboration with GSI, the INFN (Istituto Nazionale di Fisica Nucleare) is involved with an activity concerning the final tests of the accelerator QDM (Quadrupole Doublets Modules) to be installed in the SIS100 synchrotron ring. There are 81 QDM to be tested, where 54 are arc type, and the other 27 are straight type modules.

Each QDM contains two superconducting quadrupoles and up to five superconducting corrector magnets, both individually tested before their assembly in the modules. Within each QDM there is

IOP Conf. Series: Journal of Physics: Conf. Series **1226** (2019) 012017 doi:10.1088/1742-6596/1226/1/012017

also room for cold masses with ancillaries like the CL (Current Leads) and the BMP (Beam Position Monitor). Several other devices are needed to keep the temperature and vacuum of the QDM under control. Bilfinger Noell GmbH in Würzburg, a company selected by GSI-FAIR, assembles the cryostat containing all this stuff.

In this paper we first describe the laboratory facilities already present, analyzing the degree of commissioning of various parts. These will include the cryogenic system, part of the ancillary cryogenics, vacuum systems and, spare electronics. Then we summarize the test list agreed with GSI, and according to this list, we describe the equipment we are procuring to be integrated into the test facility. Part of the work in progress concerns the realization of a global control system to achieve a fully functional testing facility, with the required reliability for performing all the tests.

2. Laboratory layout and main instruments

The laboratory has its own building realized at University of Salerno in the campus area, in the southern of Italy. The laboratory spans a length of over 30 meters, a width of 15 meters, and 9 meters height, thus it has an indoor available surface of 450 square meters. Two additional side rooms, each one of about 40 square meters and 3 m ceiling, are also available. The main shed is equipped with a 20 tons crane to allow movement of heavy weight inside the building. In addition a large front gate allows the access of trucks inside.

On the backside of the shed there is an external area where technical services have been installed, like the cooling water evaporation tower, the pure helium gas storage, the helium main screw compressor of the cryogenic refrigerator. A layout of these technical services inside and outside the shed is shown on Fig. 1 where, the space around has been left available for future expansions or needs.

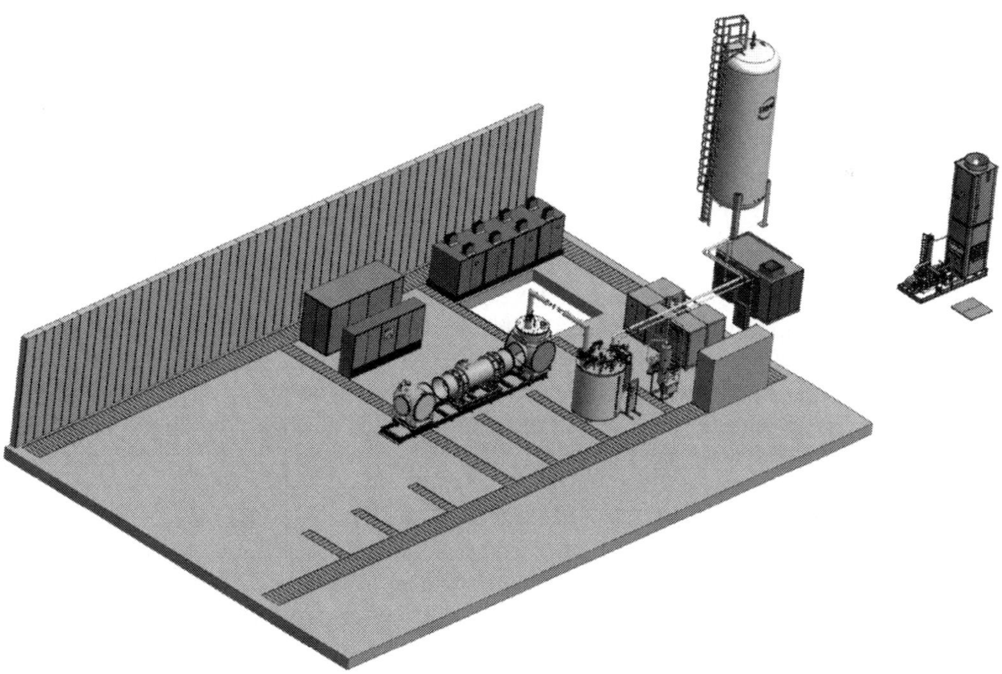

Figure 1. Layout of the laboratory with the main equipment required for testing the QDM magnet. On the right outside, there are: the cooling water evaporation tower, the tall pure helium gas storage and near the shed the main screw compressor. Inside the shed, the QDM cryostat is horizontally fixed to the refrigeration plant through a Feed-Box, connected to the Cold-Box with a cryogenic transfer line.

The total electrical power available to the laboratory is 1.2 MW and is provided by a dedicated 16 kV medium voltage substation, located near the main shed. In the future, if necessary, this power substation may be upgraded up to 2.5 MW.

In order to comply with all the GSI requests, new certified instruments have been bought and for the others, already in use, a quality certification has to be obtained.

The core of the cryogenic plant is the LR280R one of the largest refrigerator, manufactured by Linde Kryotechnik Grooup. Our configuration has been customized for present need and future expansions. The cryogenic plant is composed by the following items:

1. Supercritical He (with pressures in the range 3÷7 bar)
2. Liquefactor (with cryogenic turbine expansion box and gas purifier)
3. Cryogenic Shield flow
4. Screw Compressor (with a 196 kW SFC, Static Frequency Converter)

In particular the Linde LR280R refrigerator is equipped with an helium subcooler stage for single phase stream at 4.5 K up to 7 bar, a separate J-T liquid port, an internal purifier and a 80 K adsorber. In refrigeration mode the maximum 4.5 K helium flow is about 15 g/s, thus the refrigerator can deliver an isobaric power of 200 W (between 4.5K÷6K), plus additional 500 W (60K÷80K) in refrigeration mode, while providing 1.5 g/s of helium at 4.6K (used for cooling the current leads) and returning to the warm side at low pressure suction (*i.e.* not through the cold box internal heat exchangers). In liquefaction mode the cold box is able to produce up to 120 l/h of liquid helium. The refrigerator controller is a Linde standard rack whose management is based on Siemens software WINCC.

The cryoplant has been tested either at the Linde factory and on site. The test on site has been performed in pure refrigeration mode on a dummy load. At regime, the Cold Box was able to acheive about 220 W @ 4.5 K and 500 W @ 65 K.

The helium inventory available is in the external storage vessel, whose capacity is of 30 cubic meters at a pressure of 10 bar, corresponding to a total of 300 cubic meters of pure helium gas.

The SFC screw compressor of 196 kW can deliver about 70 g/s of helium at the pressure of 11 bar with the full speed of 1500 rpm. The SFC electronics, controls the compressor flow through its speed: at low flow (*i.e.* at low cryogenic power) the speed can be dropped up to 750 rpm, reducing the compressor input power to about 50% of the full power.

External liquid nitrogen cryoabsorber has been recently added in the helium process lines. This item allows a more efficient removal of the gas contaminant from the helium lines during the purging procedure. A dedicated Multi Component Detector gas analyser will monitor the presence of helium contaminants (mostly moisture and nitrogen) in selected places of the He lines.

The QDM module is fixed into its horizontal cryostat, and, as shown in Fig. 1, is connected to the refrigeration plant through a Feed-Box, which is connected to Cold-Box by a cryogenic transfer line. The Feed-Box is the place where hydraulic and electrical connections converge at low temperature. In fact on the top flange of the Feed-Box two current leads may be fitted, to deliver the electrical power at low temperature [1]. For the foreseen tests the high current leads are not required, so the corresponding flanges are temporary closed with blind flanges. The Feed-Box will also provide the vacuum pumping system and analysis, and will contain also the cold flow meters. In addition also an End-Box has to be realized to close the other termination of QDM cryostat. The electrical connections of all the electronic devices installed inside the module are routed on a flange placed on the center of the module. Besides the two ends and the central flange, there is no other access inside the module.

Following the GSI design and instructions, a railway system, that simplifies the alignment of the QDM cryostat with the Feed-Box and End-Box has been designed and is actually under construction.

A detailed three-dimensional model of these units is shown in Fig. 2 where, on the left of the figure there is the feed box with, on its top, the two cryogenic high current leads and the cooling pipes needed to cool down the superconducting coils. The other termination box, on the right side of the same figure, plays the role of both module mechanical support and vacuum-tight closure.

IOP Conf. Series: Journal of Physics: Conf. Series **1226** (2019) 012017 doi:10.1088/1742-6596/1226/1/012017

It has to be noticed the important role, played by the vacuum pumping groups. In fact, a very good insulation vacuum is needed in order to cool down quickly and efficiently the module under test. Actually, not shown in the previous figures, the Feed-Box is equipped with a pumping group, that make the insulation vacuum of the whole cryogenic system, including the QDM module shares which shares the same insulation vacuum of the Feed-Box.

An additional pumping group is needed for maintaining the vacuum within the beam pipe. Due to the specific needs, it is required that the beam pipe has to be maintained as clean as possible and with the highest vacuum achievable. In order to fulfill this request a specific pumping system completely oil free is installed on the End-Box with the scope of maintaining the solely vacuum of this beam pipe.

Figure 2. Scaled 3D model of the test assembly unit that will be realized for the SIS100 module, shown on the center. The frontend boxes are on both sides of the image while the rail support is shown under the quadrupole module. (Model realized by A Buonora of the Buonora Engineering Studio©)

As a final implementation of the vacuum quality control within the cryostat, there is a leak detector and two RGA (Residual Gas Analyzer) that will permits to quickly identify the origin and the nature of the residual gas and act accordingly.

3. Modules testing procedure

After the installation and before starting to work on the real modules, all the instruments and apparatus have to be designed, manufactured, and tested. The quality tests, required by the GSI for all modules, are essentially vacuum-tight, electrical and thermal both at room and at the cryogenic working temperature. The order of the test steps has been organized in such a way that an early failure may lead to an abort all the tests that will follow. However, the decision to stop or proceed in case of a test failure will be done on a case-by-case basis and, will be done in agreement with the GSI engineers in charge evaluating, among other considerations, the economical convenience in saving both time and economic resources with an early reject.

A first visual test is done at the arrival. The QDM are assembled by Bilfinger&Noell in Germany, and shipped to Salerno by truck. The QDM is accepted if do not have any visual damage and the enclosed documents contains the factory test report with all the right thermometer calibration curves.

After the inspection the QDM is mounted on the test assembly unit and all pipes and wires are connected. The pumping groups are activated and the insulation vacuum slowly restored. Within few hours large leaks can be detected, while smaller leaks require the activation of the leak detector and some helium gas to be insufflated into the module pipes. Anyway, either large or small leaks need to be found and if possible closed, until the final vacuum pressure is reached.

After the module is vacuum tight mounted, the control electronic, mainly thermometers and heaters, may be connected to the computer controlled relay box. Table 1 will summarize the device and the number of connections needed in order to verify the low voltage devices. The high voltage test points, about ten points, depending on the particular module under test, are not reported in the Table 1.

A computer controlled Relay Switchboard will manage all the wires from the connections reported on Table 1 and will connect all the termination to the instruments in order to execute the wire connection test, electrical functionality test and high or low voltage insulation tests.

N. Devices	Device description	Terminals	Connections	Notes
20	Cernox/CCS CLTS	4	80	Only wires and readings
13	PT100 Resistors	2	26	Wires and Resistance readings
5	Thermocouples	2	10	Only Voltage readings
18	Heaters	2	36	Power test and thermal coupling
1	Cryocatcher	1	1	Only Insulation
Total number of required connections			153	

Table 1. List of the devices present inside one of the cryostat module, mainly thermometers and heaters for monitoring and temperature control. The connections column reports the total number of wires needed. The last column reports the kind of tests that can be done on that type of device.

The wire connection tests apply to the four-contact thermometers, where it is not possible to prove the device, usually a doped semiconductor with unknown resistance. The test measure the electrical resistance of the two couple of wires joined together on the same sensor terminal ensuring that both couples of wires are connected together on their extremity. On the thermocouples, with only two-contact connections and negligible intrinsic resistance, this measure is applied to the two contacts. The test may be considered passed if the measured resistance values are compatible with the expected value of resistance of the wires. The connection test on the two-contact heaters and PT100 sensors is the same as thermocouples. In this case, the test may be considered passed if the measured value is compatible with the sum of the resistance of the wires and the sensor.

The electrical functionality test on the thermometers is obtained with the reading the temperature values. At room temperature, in isothermal condition, within the variation of about a degree, all thermometers should report the room temperature value. After the cool down, the same measure is less meaningful and may be omitted. The heater functionality requires a thermal coupling test. This test may be considered passed if after pouring some power into the heater, the corresponding thermometer shows a significant rise in its temperature value.

The high voltage insulation tests apply only to the superconducting taps on coils and bus-bars [2]. The tests must be done above the Dew Point with a temperature below 40 °C. Three voltage steps of 200, 500 and 1000 Volts, with respect to ground, are applied on each voltage tap for 120 s with a 120 s discharge pause before and between them [3]. Alternatively, a voltage ladder with five linear steps from 200 to 1000 Volts with 120 s duration of each step may be used. A time interval of 240 s for discharge is required before reversing the polarity and at the end of the test. If while increasing the voltage the resistance falls below 100 Mega Ohm the test will be interrupted. The test is considered passed if the insulation resistance is greater than 1 Giga Ohm for any voltage step.

The low voltage insulation tests, apply to the electronic devices, mainly thermometers and heaters by measuring the leakage current to ground. In order to protect the low voltage device, during the measure, all terminal have to be joined together. Two voltage steps of 20 and 100 Volts with respect to ground are applied, each for 120 s, with both polarities. This test is considered passed if the resistance toward the ground is higher than 100 Mega Ohm.

During the execution of all tests, room temperature and humidity are recorded together with the atmospheric pressure. While pumping down the cryostat the value of the vacuum is monitored and the spectra of residual gases recorded. Resistances and leakage currents will be acquired during the whole duration of all insulation tests with a rate of 2 sample/s. All graphs will be added to the final report.

After the room temperature tests, the cooling helium gas provided by the Linde LR280R refrigerator will chill the superconducting quadrupole within the cryostat from room temperature to about 4.5 K. During the cool down of the QDM, besides the control of the residual vacuum pressure, and temperatures, there will be a careful registration of some relevant signals extracted from the Cold Box. The mass flow and the differential pressure of the in/out cooling helium gas will be measured and monitored. The module will be considered cooled down when the designed working cryogenic temperatures are reached and carefully stabilized.

While the QDM is kept at this temperature value, the room temperature electrical tests have to be repeated almost identically. In this state, the tests of the superconducting sextupoles now may be executed. The coils will be powered with a current ramp, with increasing value of intensity and rate, up to the working values are reached and slightly overcome. During this test, a highly sensitive quench detector, inside the power supply, will work in order to avoid damages to the superconducting coils and to the high-temperature superconducting current leads [4].

At the end of all cryogenic temperature electrical tests, the module under test, will naturally return toward room temperature. Because of the large masses involved this may require several days. The circulation of warm helium gas, within the cooling pipes, may significantly accelerate the warming up process. The control of the helium flux and its temperature, may significantly and safely shorter the warming times. When, finally, the module is warmed up and stabilized at room temperature, the room temperature electrical tests are repeated again. This test repetition will ensure that the thermal cycle has not caused damages to the module coils and on the devices within it.

Through all the tests the collected data are stored locally. The decision of accepting or rejecting the tested module will be done after the evaluation of the information contained in this file. A dedicated Report Generator program will construct the final report by using the log and data files, written during the test procedures and monitoring activities. All log files and data file, as soon as they are created, will be stored in a folder as text files in order to be always readable on any computer. While reading the information contained in the log files, the Report Generator will control the acquired values and create the required graphics. Finally, data, text, and pictures will be merged into a single unalterable pdf document. The log and data files used by the Report Generator program, in order to save their integrity, will be compressed into a single password-protected archive and stored in multiple copies into dedicated storage sites.

4. Final remarks and conclusions

The INFN group within the University of Salerno is involved into the setting up of a laboratory for testing part of the superconducting magnets to be installed in the SIS100 synchrotron ring.

IOP Conf. Series: Journal of Physics: Conf. Series **1226** (2019) 012017 doi:10.1088/1742-6596/1226/1/012017

Since the tests are required on about eighty almost identical modules, a testing procedure has been created with the highest possible automation level. This will reduce the operator errors and will speed up the time needed in some repetitive measurements. This procedure at its end will generate a report that constitutes the most important document on which the decision of accepting or rejecting the quadrupole module will be taken. In this paper the procedure of quality verification of the quadrupole modules is presented in some details with a description on how the proposed steps will operate.

Nowadays this type of facility is not yet present in Italy and in Europe only few, very specialized research centers, are able to carry out testing procedure of such complexity with the needed expertise. The buildup of this laboratory for the complexity and the efforts required and the technological breakthrough represents a challenge for the Italian physics community. The presence of this scientific facility would also contribute to expand the competence of firms located on the surroundings.

References

[1] Saggese A, Iannone G, Gambardella U, Califano N and Ferrentino A 2015 *IEEE Trans. On Applied Supercond.* **25** 4801304.

[2] Rahimzadeh-Kalaleh S, Ambrosio G, Chlachidze G, Donnelly C and Tartaglia M 2009 *IEEE Trans. On Applied Supercond.* **19** 2442-2445.

[3] Zhang Z, Song Y, Wu H, Xie Y, Yang Z, Shen G, Wu W, Lu K and Wei J 2017 *IEEE Trans. On Applied Supercond.* **27** 4900804.

[4] Kamby P and Elkiær A 2012 Test Procedure 501717-201 4x855 20kA/±25V and 10kA/±50V Danfysik Factory.

IOP Conf. Series: Journal of Physics: Conf. Series **1226** (2019) 012018 doi:10.1088/1742-6596/1226/1/012018

Two-dimensional effects in Fowler-Nordheim field emission from transition metal dichalcogenides

F Urban[1,3], M Passacantando[2,3], F Giubileo[3], L Iemmo[1,3], G Luongo[1,3], A Grillo[1], A Di Bartolomeo[1,3]*

[1] Department of Physics "E.R. Caianiello", University of Salerno, 84084, Fisciano, Italy;
[2] Department of Physical and Chemical Science, University of L'Aquila, and CNR-SPIN L'Aquila, via Vetoio, Coppito 67100, L'Aquila, Italy;
[3] CNR-SPIN Salerno, 84084, Fisciano, Italy;
* Correspondence: adibartolomeo@unisa.it; Tel.: +39-089-969-189

Abstract. We report field emission from bilayer MoS_2 and monolayer WSe_2 synthesized by CVD on SiO_2/Si substrate. We show that the emitted current follows a Fowler-Nordheim model modified to account for the two-dimensional confinement of charge carriers. We derive the figures of merit of field emission and demonstrate that few-layer transition-metal dichalcogenides are suitable for field emission applications.

1. Introduction

Transition - metal dichalcogenides (TMDs), chemical formula $\boldsymbol{MX_2}$, have been largely investigated during the last few years. They present a layered structure, characterized by a honeycomb lattice where each transition metal $\boldsymbol{M}$ $(\boldsymbol{Mo}, \boldsymbol{W}, \boldsymbol{We}, \textbf{etc.})$ is covalently bonded to two chalcogenide atoms $\boldsymbol{X}$ $(\boldsymbol{S}, \boldsymbol{Se}, \boldsymbol{Te})$. TMDs have been intensively studied for the ease of fabrication and the tuneable and sizeable bandgap suitable for optoelectronic applications [1,2]. Like graphene, TMDs can be mechanically exfoliated from a bulk material and transferred on a substrate [3]; however, single-crystal and large-scale flakes, with controlled number of layers, are more easily produced by chemical vapor deposition (CVD) [4]. Those materials are considered good candidates for technological applications because they offer good control on of the energy bandgap through the number of layers [5], which increases passing from bulk to monolayer form, high on/off ratio ($> 10^7$) [6] and low subthreshold swing (below $70\ \mathrm{mV/decade}$) [7,8], even though their mobility is typically only few tens $\mathrm{cm^2V^{-1}s^{-1}}$[9,10]. Their strong light absorption and emission open the route for high-responsivity photodetectors, high-intensity light emitters and efficient photovoltaic cells [11,12]. Furthermore, the intrinsic sharp edges due to the 2D nature and the low work function enable field emission applications. Anyway, unlike graphene and carbon nanotubes (CNTs) [13,14], TMDs have not been much explored as cold cathodes for applications in vacuum electronics, flat panel displays, electron microscopy, x-ray tubes [15–17], etc. Only few studies focussed on field emission from MoS_2 and WSe_2 have been reported to date [18–20]. In this paper, we investigate the field emission (FE) behaviour of MoS_2 and WSe_2 bilayer and monolayer flakes. Specifically, we report FE measurements from their flat part. We show that the experimental data are well fitted by a Fowler-Nordheim (FN) model [21] modified to take into account the two-dimensional (2D) nature of the material. We derive the characteristic FE parameters such as the field enhancement factor β and the turn on field to show that TMDs are promising materials for field emission applications.

2. Materials and Methods

We grew the two analysed samples, $\boldsymbol{MoS_2}$ and $\boldsymbol{WSe_2}$ flakes, by chemical vapor deposition on a highly p-doped $\boldsymbol{Si}$ substrates covered by $\mathbf{300}\ \boldsymbol{nm}$ of $\boldsymbol{SiO_2}$ (silicon dioxide). The CVD growths, at 850 °C, were performed in a two-zone quartz tube furnace, with the substrate placed face down on the transition-metal precursor and the chalcogen carried by an Ar gas flow [22]. Flakes were randomly formed on the substrates. Figure 1(a) and Figure 1(b) show SEM images of two selected flakes, $\boldsymbol{MoS_2}$ and $\boldsymbol{WSe_2}$, respectively. The field emission measurements were performed in the SEM chamber, at a pressure $< \mathbf{10^{-6}}\ \boldsymbol{mbar}$, exploiting the piezo-controlled tungsten tip ($\boldsymbol{W}$-tip) movement for the accurate positioning of the anode in front of the emitting flake (cathode) at a close distance $\boldsymbol{d}$, as displayed in insets of Figure 1(a) and (b).

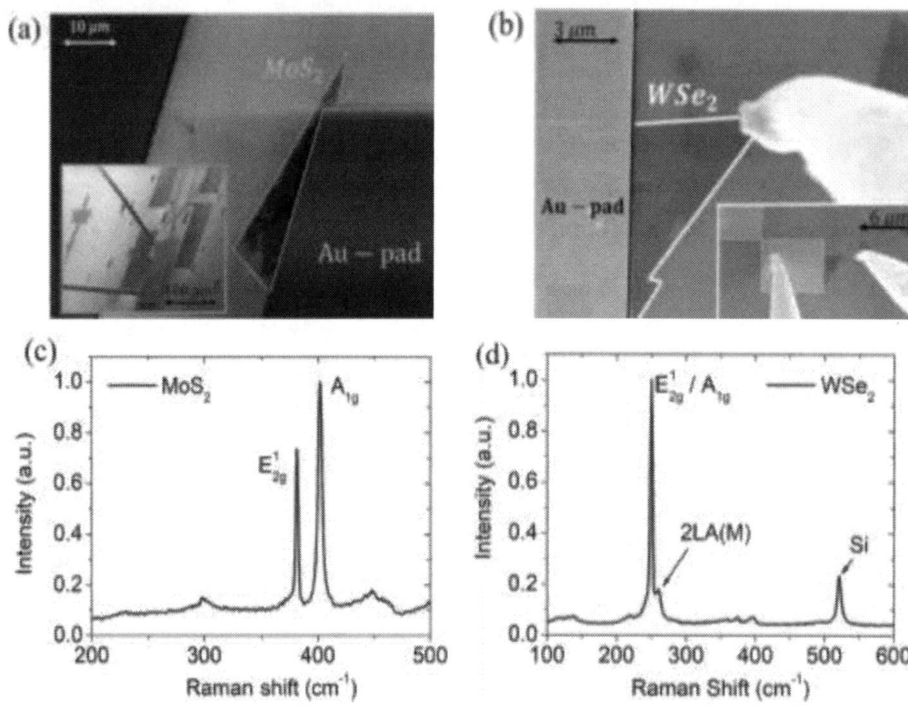

Figure 1. SEM images of the flakes used for field emission measurements and in insets the position of the tungsten tips: the lower one on the metal pad contacting the flake, the other placed in front of the flake at a close distance *d*, respectively for MoS_2(a) and WSe_2(b). Raman spectra of the MoS_2(c) and WSe_2(d).

The flakes were characterized by Raman spectroscopy with an unpolarized incident laser to identify the number of layers. The Raman spectrum reported in Figure 1(c), referring to MoS_2, shows two peaks at $\sim 381\ cm^{-1}$ and $\sim 401\ cm^{-1}$ which are associated with the in-plane vibrations of sulfur atoms (E^1_{2g}) and with the out-of-plane vibrations of Mo and S atoms (A_{1g}), respectively [23]. The separation of $\sim 20\ cm^{-1}$ indicates bilayer flakes. Similarly, Figure 1(d) shows the Raman spectrum of the WSe_2 flake. It exhibits two peaks around $\sim 250\ cm^{-1}$ and $\sim 260\ cm^{-1}$, corresponding the

IOP Conf. Series: Journal of Physics: Conf. Series **1226** (2019) 012018 doi:10.1088/1742-6596/1226/1/012018

first to a combination of the in-plane vibrations of W and Se atoms (E_{2g}) and out-of-plane vibrations of Se atoms (A_{1g}), and the other one to a second order resonant Raman mode ($2\,LA\,(M)$) due to LA phonons at the M point in the Brillouin zone [24]. The peak frequency positions are compatible with WSe_2 monolayer, which was further confirmed by photoluminescence measurements (not shown here) [18].

3. Results

Considering that the MoS_2 and WSe_2 flakes are naturally n-doped and that the sharp edges or nano-protrusions of the flat-part can originate a high electric field amplification [14,25], we expect easy extraction of electrons upon application of a voltage.

According to the standard FN theory [26] the field emission current usually results in a linear $\ln(I_{ds}/(V_{ds}^2))$ vs. $1/V_{ds}$ FN plot. However, for 2D materials, it has been proposed [21] that a modified FN model has to be applied to the vertical electron emission to take into account the reduced dimensionality, the energy spectrum, the non-conserving in-plane electron momentum, the finite-temperature and the space-charge-limited effects.

The modified FN model results in the following field emission current

$$I_{FN}^{2D} = A_{FN}^{2D} \exp\left[-\frac{b\Phi^{\frac{3}{2}}}{\beta E}\right] \quad (1)$$

where A_{FN}^{2D} and b are constants, S and ϕ represent the emitting surface and the electron affinity (in our case $\phi = 4.2\,eV$ for MoS_2 [27] and $3.9\,eV$ for WSe_2 [28]), β is the so-called field enhancement factor, a typical figure of merit for the qualification of field emitter material. Finally, the electric field is $E = V/d$, where V represents the anode-cathode voltage and d their distance.

For both samples the FE current was measured with the (anode) W-tip positioned at a fixed distance ($d \sim 200\,nm$ for MoS_2 and $d \sim 400\,nm$ for WSe_2) from the surface (inner flat part of the flake). We disregarded the edges after having verified that the oxidation after air exposure and the higher concentration of unreacted precursor (MO_x and WO_x) had hampered the extraction of electrons from them. The current was monitored during the voltage sweep up to 120 V (a constraint to avoid the oxide breakdown).

Figure 2(a) and 2(b) show that, when V is below a critical value, the current fluctuates around the noise floor $< 10^{-13} A$, and there is no measurable signal. For $V_{ds} > V_{th}$ an exponential increase (well visible in insets of Figure 2(a) and (b)) in the current is observed. The current-voltage characteristics show the typical fluctuations of emission current, related to desorption of adsorbates caused by Joule heating.

The fits of the experimental data at high-field regime of eq.1 are shown in Figure 2(c) and 2(d) compared with the 3D Fowler-Nordheim model [26], reported in inset. The accuracy of the linear fit is evaluated with the $\chi^2 \sim 0.99$, defining a good agreement in a wide range of voltage values, slightly better for WSe_2 respect to the classical 3D Fowler-Nordheim model.

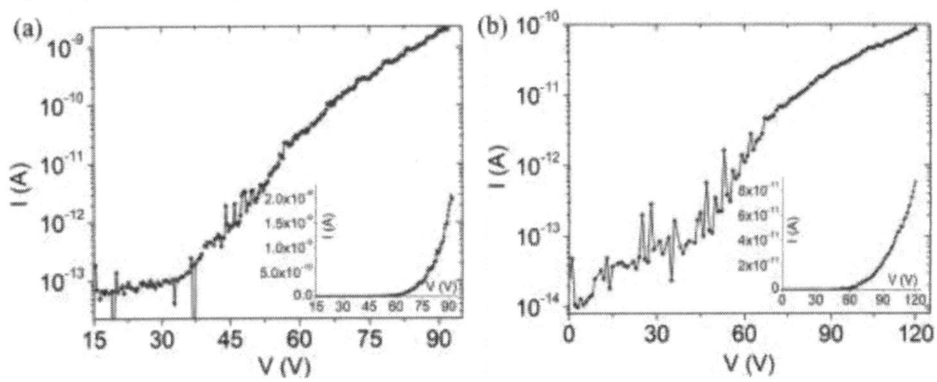

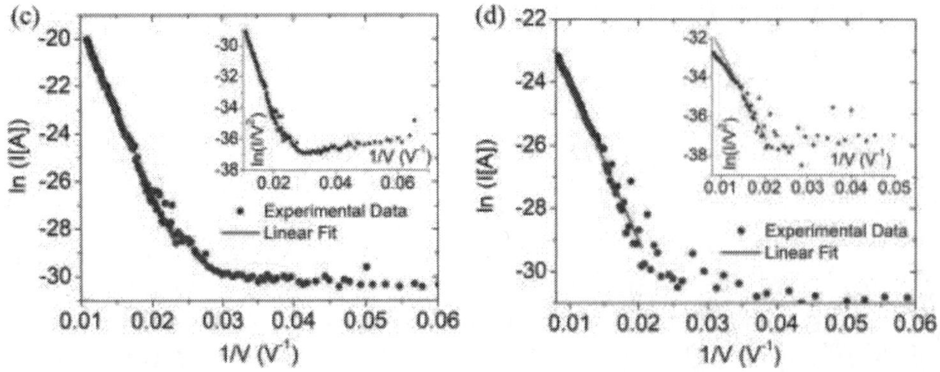

Figure 2. Field emission curves plotted in logarithmic scale as a function of anode-cathode voltage and linear scale (inset) for MoS_2(a) and WSe_2(b). Experimental data (blue circles) and model fit (red lines) of field emission measures for MoS_2(c) and WSe_2(d). The insets show the comparison of the experimental data fitted with standard Fowler-Nordheim model.

Indeed, this provides an indication in favour of the 2D modified FN model proposed in ref. [21], that should be further addressed and confirmed in future experiments.
From the slopes of the linear fits (eq.1) we derive the field enhancement factor $\beta = -b\ d\ \phi^{3/2}/m \sim 17$ and ~ 40 for MoS_2 and WSe_2 respectively.
The seemingly low values of the amplification factor (good emitters reach $\beta \sim 1000$ or more) are remarkable if we consider that emission happens from the inner part of the flake, without taking advantage from the 2D shape of the flake. Despite that, the turn-on field, defined as the field applied to extract a current of 1 pA has a value of $\sim 230\ V/\mu m$ and $150\ V/\mu m$ for MoS_2 and WSe_2, respectively, which are rather low compared to the typical turn on field of several $kV/\mu m$ needed to extract electrons from flat surfaces. These results demonstrate the MoS_2 and WSe_2 suitability as field emitters.

4. Conclusion

In conclusion, we have fabricated MoS_2 bilayers and WSe_2 monolayer flakes and analysed their field emission properties. We have demonstrated that a modified field emission model, which considers the 2D nature of the flakes, provides a slightly better agreement with the experimental data compared to the original Fowler-Nordheim model, opening the route for further investigations. We have highlighted that TMDs are promising materials not only for electronics and optoelectronics but also for field emission applications.

References

[1] Di Bartolomeo A, Genovese L, Giubileo F, Iemmo L, Luongo G, Tobias Foller and Schleberger M 2018 Hysteresis in the transfer characteristics of MoS 2 transistors *2D Materials* **5** 015014

[2] Pospischil A and Mueller T 2016 Optoelectronic Devices Based on Atomically Thin Transition Metal Dichalcogenides *Applied Sciences* **6**

[3] Lin Z, McCreary A, Briggs N, Subramanian S, Zhang K, Sun Y, Li X, Borys N J, Yuan H, Fullerton-Shirey S K, Chernikov A, Zhao H, McDonnell S, Lindenberg A M, Xiao K, LeRoy B J, Drndić M, Hwang J C M, Park J, Manish Chhowalla, Schaak R E, Javey A, Hersam M C, Robinson J and Terrones M 2016 2D materials advances: from large scale synthesis and controlled heterostructures to improved characterization techniques, defects and applications *2D Materials* **3** 042001

[4] Feng Y, Zhang K, Wang F, Liu Z, Fang M, Cao R, Miao Y, Yang Z, Mi W, Han Y, Song Z and Wong H S P 2015 Synthesis of Large-Area Highly Crystalline Monolayer Molybdenum

Disulfide with Tunable Grain Size in a H2 Atmosphere *ACS Applied Materials & Interfaces* **7** 22587–93

[5] Mak K F, Lee C, Hone J, Shan J and Heinz T F 2010 Atomically Thin MoS2: A New Direct-Gap Semiconductor *Phys. Rev. Lett.* **105** 136805

[6] Zhou C, Wang X, Raju S, Lin Z, Villaroman D, Huang B, Chan H L-W, Chan M and Chai Y 2015 Low voltage and high ON/OFF ratio field-effect transistors based on CVD MoS2 and ultra high-k gate dielectric PZT *Nanoscale* **7** 8695–700

[7] Nourbakhsh A, Zubair A, Joglekar S, Dresselhaus M and Palacios T 2017 Subthreshold swing improvement in MoS2 transistors by the negative-capacitance effect in a ferroelectric Al-doped-HfO2/HfO2 gate dielectric stack *Nanoscale* **9** 6122–7

[8] Fang H, Chuang S, Chang T C, Takei K, Takahashi T and Javey A 2012 High-Performance Single Layered WSe2 p-FETs with Chemically Doped Contacts *Nano Lett.* **12** 3788–92

[9] Lopez-Sanchez O, Lembke D, Kayci M, Radenovic A and Kis A 2013 Ultrasensitive photodetectors based on monolayer MoS2 *Nature Nanotechnology* **8** 497

[10] Zhou H, Wang C, Shaw J C, Cheng R, Chen Y, Huang X, Liu Y, Weiss N O, Lin Z, Huang Y and Duan X 2015 Large Area Growth and Electrical Properties of p-Type WSe2 Atomic Layers *Nano Lett.* **15** 709–13

[11] Zheng Z, Zhang T, Yao J, Zhang Y, Xu J and Yang G 2016 Flexible, transparent and ultra-broadband photodetector based on large-area WSe 2 film for wearable devices *Nanotechnology* **27** 225501

[12] Di Bartolomeo A, Genovese L, Foller T, Giubileo F, Luongo G, Luca Croin, Liang S-J, Ang L K and Schleberger M 2017 Electrical transport and persistent photoconductivity in monolayer MoS 2 phototransistors *Nanotechnology* **28** 214002

[13] Di Bartolomeo A, Giubileo F, Iemmo L, Romeo F, Russo S, Unal S, Passacantando M, Grossi V and Cucolo A M 2016 Leakage and field emission in side-gate graphene field effect transistors *Appl. Phys. Lett.* **109** 023510

[14] Iemmo L, Di Bartolomeo A, Giubileo F, Luongo G, Passacantando M, Niu G, Hatami F, Skibitzki O and Schroeder T 2017 Graphene enhanced field emission from InP nanocrystals *Nanotechnology* **28** 495705

[15] Giubileo F, Di Bartolomeo A, Iemmo L, Luongo G, Passacantando M, Koivusalo E, Hakkarainen T V and Guina M 2017 Field Emission from Self-Catalyzed GaAs Nanowires *Nanomaterials* **7**

[16] Yabushita R, Hata K, Sato H and Saito Y 2007 Development of compact field emission scanning electron microscope equipped with multiwalled carbon nanotube bundle cathode *Journal of Vacuum Science & Technology B: Microelectronics and Nanometer Structures Processing, Measurement, and Phenomena* **25** 640–2

[17] Gupta A P, Park S, Yeo S J, Jung J, Cho C, Paik S H, Park H, Cho Y C, Kim S H, Shin J H, Ahn J S and Ryu J 2017 Direct Synthesis of Carbon Nanotube Field Emitters on Metal Substrate for Open-Type X-ray Source in Medical Imaging *Materials* **10** 878

[18] Di Bartolomeo A, Urban F, Passacantando M, McEvoy N, Peters L, Iemmo L, Luongo G, Romeo F and Giubileo F 2018 A WSe2 vertical field emission transistor *ArXiv e-prints*

[19] Urban F, Passacantando M, Giubileo F, Iemmo L and Di Bartolomeo A 2018 Transport and Field Emission Properties of MoS2 Bilayers *Nanomaterials* **8**

[20] Gaur A P S, Sahoo S, Mendoza F, Rivera A M, Kumar M, Dash S P, Morell G and Katiyar R S 2016 Cold cathode emission studies on topographically modified few layer and single layer MoS2 films *Applied Physics Letters* **108** 043103

[21] Sin Ang Y, Zubair M, Ooi K J A and Ang L K 2017 Generalized Fowler-Nordheim field-induced vertical electron emission model for two-dimensional materials *arXiv:1711.05898*

[22] O'Brien M, McEvoy N, Hallam T, Kim H-Y, Berner N C, Hanlon D, Lee K, Coleman J N and Duesberg G S 2014 Transition Metal Dichalcogenide Growth via Close Proximity Precursor Supply *Scientific Reports* **4** 7374

[23] Li X and Zhu H 2015 Two-dimensional MoS2: Properties, preparation, and applications *Journal of Materiomics* **1** 33–44

[24] O'Brien M, McEvoy N, Hanlon D, Hallam T, Coleman J N and Duesberg G S 2016 Mapping of Low-Frequency Raman Modes in CVD-Grown Transition Metal Dichalcogenides: Layer Number, Stacking Orientation and Resonant Effects *Scientific Reports* **6** 19476

[25] Di Bartolomeo A, Passacantando M, Niu G, Schlykow V, Lupina G, Giubileo F and Schroeder T 2016 Observation of field emission from GeSn nanoparticles epitaxially grown on silicon nanopillar arrays *Nanotechnology* **27** 485707

[26] Fowler R H and Nordheim L 1928 Electron emission in intense electric fields *Proc R Soc London A* **119** 173–81

[27] Choi S, Shaolin Z and Yang W 2014 Layer-number-dependent work function of MoS2 nanoflakes *Journal of the Korean Physical Society* **64** 1550–1555

[28] Smyth C M, Addou R, McDonnell S, Hinkle C L and Wallace R M 2017 WSe 2 -contact metal interface chemistry and band alignment under high vacuum and ultra high vacuum deposition conditions *2D Materials* **4** 025084

IOP Conf. Series: Journal of Physics: Conf. Series **1226** (2019) 012019 doi:10.1088/1742-6596/1226/1/012019

Thermal compensation system in advanced and third generation gravitational wave interferometric detectors

L Aiello[1,2], E Cesarini[3], V Fafone[1,4,5], M Lorenzini[1,2], Y Minenkov[5], I Nardecchia[1,2], A Rocchi[5] and V Sequino[4,5]

[1] Gran Sasso Science Institute, Viale Francesco Crispi 7, I-67100, L'Aquila, Italy
[2] INFN - Laboratori Nazionali del Gran Sasso, Via Giovanni Acitelli 22, I-67100 Assergi, Italy
[3] Centro studi e ricerche Enrico Fermi, Piazza del Viminale 1, I-00184, Roma, Italy
[4] Dipartimento di Fisica, Università di Roma Tor Vergata, Via della Ricerca Scientifica 1, I-00133, Roma, Italy
[5] INFN - Roma Tor Vergata, Via della Ricerca Scientifica 1, I-00133, Roma, Italy

E-mail: `lorenzo.aiello@gssi.infn.it`

Abstract. On September 14th 2015 the first gravitational wave signal has been detected by the Advanced LIGO interferometers, opening the era of the gravitational astronomy and giving new opportunities to investigate the universe. The Advanced LIGO and Advanced Virgo interferometers are now back in a commissioning phase in order to improve their sensitivity for the next observing run, which will start in the first months of 2019.
In the high-frequencies region of their sensitivity band, the detectors are shot-noise limited: the sensitivity in this frequency window could be improved increasing the laser input power, but this increases also the optical aberrations due to the thermal effects. The optical power absorptions in the substrate and coatings of the optics induce both an increase of the optical path length in the substrates of the mirrors (thermal lensing) and a thermal expansion of the optic itself along the optical axis (thermo-elastic deformation). Both these aberrations reduce the sensitivity of the detector, limiting its performances. In order to face and minimize them, an adaptive Thermal Compensation System is required in order to guarantee the proper operation of the interferometer.
An overview of the present Thermal Compensation System system installed on Advanced Virgo, with also a focus on the possible improvements of the actual actuators for the next generation of detectors, is presented here.

1. Introduction

One century after their prediction by General Relativity, Gravitational Waves have been detected: after the first detection ever made in 2015 [1] other events followed [2–5], among which the famous GW170817 signal [6, 7] emitted by the merger of two neutron stars, that marked the begin of the the multi-messenger astronomy. After the end of the O2 observing run in August 2017, the Advanced LIGO and Advanced Virgo detectors are back in a commissioning phase with the installation of new upgrades in order to further improve their sensitivity.
In order to reduce the shot noise contribution in the high frequency window of the detector band, an increase of the interferometer input power is planned for the next observing run O3. However,

the higher power circulating inside the detector will increase also the optical aberrations due to thermal effects. These aberrations are originated by the optical power absorbed in the substrate and in the coatings of the mirrors, that induce both an increase of the optical path length in the substrates of the mirrors (*thermal lensing*) and a thermal expansion of the optic itself along the optical axis (*thermo-elastic deformation*). As a consequence, some light is scattered from the fundamental mode to higher-order ones: the error signals to maintain the cavities at resonance become fainter, resulting in a loss of robustness and sensitivity that reduces the performances of the detector. In order to face and minimize the optical aberrations, an adaptive Thermal Compensation System (TCS) to guarantee the proper operation of the interferometer is required.

2. The Thermal Compensation System

The TCS has been designed [8, 9] to reduce the thermal effects arising from the high power circulating in the detector. The subsystem is composed by sensors to detect the aberrations and by actuators to compensate them. The scheme of the different TCS components integrated in Advanced Virgo is shown in Figure 1. As it possible to see from it, TCS is made up of:

- Hartmann Wavefront Sensors (HWSs) dedicated to the measurement of the thermal lensing (HWS-RC) and of the thermo-elastic deformation (HWS-HR), see section 2.1;
- CO_2 laser projector and Ring Heater (RHs) for the correction of the above-mentioned optical aberration, see section 2.2.

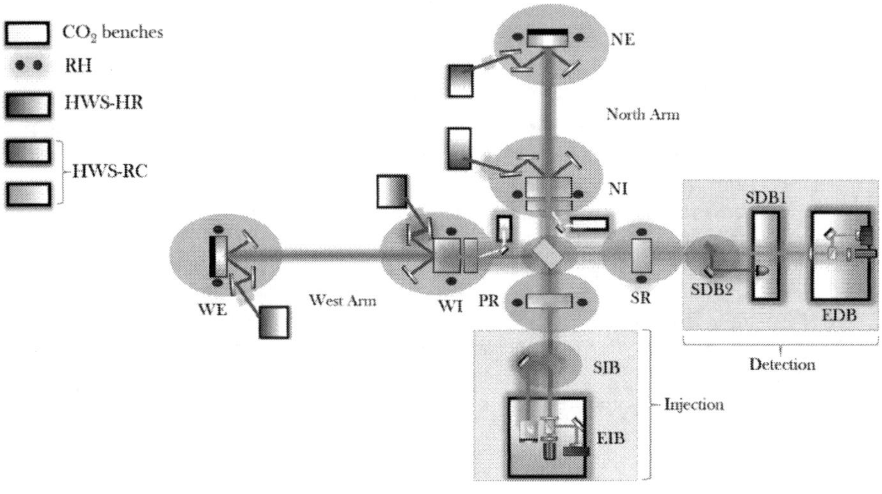

Figure 1. Scheme of the Thermal Compensation System of Advanced Virgo, superimposed on the optical layout of the detector.

2.1. *Sensing: Hartmann Wavefront Sensor*

The sensing part of the TCS is based on HWSs [10]. HWS is a differential sensor, that measures the change of a live wavefront with respect to a reference one through an uncoherent probe beam. The latter is generated using a fiber-coupled SLED (Superluminescent Diode) source: it has been demonstrated that its low spatial coherence improves the noise performance of the HWS, because eliminates interference between stray beams and reduces the cross-talk between neighboring spots [9, 10]. HWS consists of an holed plate (Hartmann plate) placed at a distance

L from a CCD camera. When the wavefront has no aberration, the SLED beam crosses the holed plate straight resulting in a series of equally distributed light spots on the CCD. The presence of aberration in the analyzed wavefront is identified from the displacement of the light spots on the CCD from the reference position. The wavefront reconstruction is then obtained by integration of the gradient field by the HWS.
In order to monitor the thermal lensing effect and the thermoelastic deformation, two different setup are used, as shown in Figure 2.

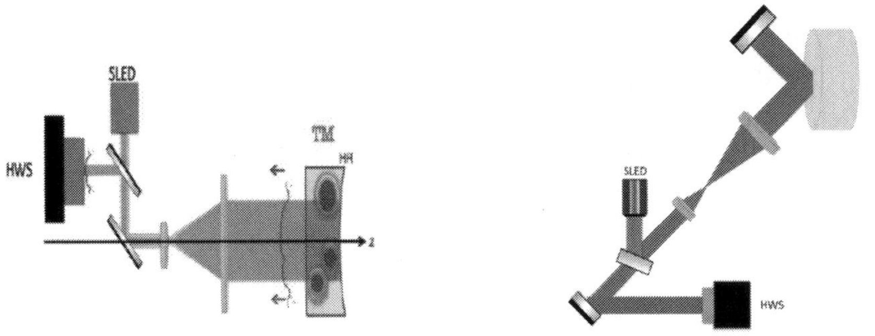

Figure 2. HWS setup for the estimation of the thermal lensing (left, on-axis measurement) and of the thermoelastic deformation (right, off-axis measurement).

For the thermal lensing estimation on the mirrors at the input of the cavities (Input Test Masses, ITMs) the SLED, superimposed on the main laser, arrives on the ITMs, pass through them and then it is reflected back toward the HWS along the same path. These are called *on-axis measurements*. Regarding the thermoelastic deformation (*off-axis measurements*), the probe beam impinges on the mirrors with an angle of 45°, then propagates up to an optics that reflects it back on the High Reflectivity surface and then toward the sensor.
Both in the on-axis and the off-axis measurements the probe beam passes on the interested area twice, making these measurements double-pass with a consequent increase of the signal-to-noise ratio.

2.2. Actuation: Ring Heater and CO_2 Laser Projector

One of the possible way to correct the thermoelastic deformation of the mirrors is to engage the RHs [11, 12]. The heaters are made of two rings of pyrex surrounded by a polished copper shield, heated by Joule effect with a wrapping of NiCr flat wire. In each RH, equipped with a pair of pyrex rings, the current in one ring flows in the opposite direction with respect to the other one: this allows to make the coupling of the stray magnetic field along the mirrors axis with actuation magnets negligible.
At the maximum power (24 W for each ring) the RH is able to reduce the radius of curvature of the mirror of 100 m over a nominal value of around 1.5 km.
Another type of actuators engaged to contrast thermal effects are 50 W CO_2 lasers. Their actuation is performed not directly on the mirrors, as it was done in the first generation of detectors: this is due to the fact that the intensity fluctuations of the CO_2 lasers could inject displacement noise into the interferometer. In the initial detectors intensity stabilization servos have been implemented and they have been able to reduce the laser fluctuations [13] but, unfortunately, the present technology is not enough developed to reduce the displacement noise up to a value below than the advanced interferometers sensitivity requirements [14]. To solve this problem the CO_2 beam, properly shaped, is shined on an additional transmissive optic, the Compensation Plate

(CP), made by fused silica and placed immediately before the ITMs. The choice of these lasers is because they emits at λ=10.6 μm, a wavelength which is strongly absorbed in the immediate surfaces of this material [15].
The optical benches on which the CO_2 laser are installed have been designed in order to allow to the lasers to shine on the CPs three different (and independent) types of actuation pattern. The first of these patterns is the *Central Heating* (CH), consisting in a CO_2 Gaussian beam shined in the CP center. As the RHs, its aim is to correct the thermo-elastic deformation but acting in the opposite way. Indeed while the RHs, heating the peripheral of the mirrors, decrease their radius of curvature, the CH is able to rise it.
The second actuation pattern is used to correct the thermal lensing due to coating absorption in the ITMs and consist in an axi-symmetric heat distribution. It is provided by two annuli-shaped intensity patterns, each of them obtained with an axicon[1] lens. The power in the two annuli can be separately regulated with dedicated waveplates, one for each annulus, and a lens placed before each axicon is used to modify the annulus thickness. This setup is called *Double Axicon System* (DAS).
The last pattern that could be engaged is the *Scanning System* (SS), that provides a correction for the residual not axi-symmetric aberrations. It consist of a CO_2 beam that arrives on two galvanometer mirrors, that move the laser beam on the surface of the CP scanning it along a 25x25 raster pattern. Power on the SS is regulated through an Acusto-Optic Modulator (AOM), and the needed intensity pattern is obtained as the sum of all the spots.

3. Improving the actuation

3.1. Mode cleaner cavity for CO_2 lasers

In order to have a good compensation, the actuators should be able to modify with high precision the shape of the laser beam: this could result quite difficult when the CO_2 lasers exhibit a not so good beam shape. To improve their actuation power, the higher-order modes must be removed from the laser beam. This could be done with a spatial beam filtering, but the efficiency of this type of system is not enough for the next generation of detectors: a mode cleaner cavity for CO_2 lasers is then needed. A design of a prototype cavity has been developed and a triangular configuration has been adopted, as shown in Figure 3.

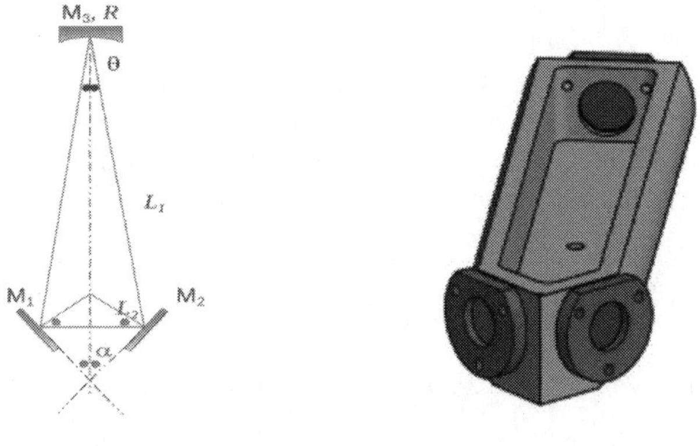

Figure 3. Conceptual layout of a three-mirror triangular mode cleaner cavity (left) and preliminary design of the studied cavity (right).

[1] The axicon is a particular type of lens with a conical surface that transform a Gaussian laser beam in an annular one.

The factor that quantifies the selectivity of the mode cleaner is the so-called *Finesse* $\mathcal{F}$, defined as its free spectral range divided by the (full width at half-maximum) bandwidth of its resonances. However, higher is the finesse, more complicated is to keep the cavity at resonance. In order to make a trade-off between the improvement of the laser beam quality and the difficulty to make the cavity work, after some calculations a reasonable value of $\mathcal{F} \sim 100$ has been chosen. Moreover, in order to allow an easy installation and integration in the actual TCS setup on Advanced Virgo, the cavity has been designed also to be compact: the total round trip of the laser inside it is less than one meter.
The common way to keep a cavity at resonance is the Pound-Drever-Hall technique [16], based on the modulation of the laser beam with an Electro-Optic Modulator (EOM). However, because of the EOM crystals for the CO_2 lasers wavelength from usual companies are too much expensive, an investigation to keep the cavity resonant with an AOM frequency shifter is in progress.

3.2. Deformable mirror

The scan and step frequency of the SS are within the sensitivity band of Advanced Virgo, then it can introduce displacement noise in the detector at the frequency of the pattern repetition rate. For this reason, it is necessary to develop new methods to shape the CO_2 laser beam: one of the most promising makes use of Micro-Electro Mechanical-System (MEMS) deformable mirrors (DM), which imprint a phase modulation to the wavefront of the incoming Gaussian beam that converts the beams intensity distribution into the desired one holographically [17]. Currently a DM [18] is under test in Tor Vergata University: the goal of the ongoing tests is to evaluate the influence functions of its actuators, i.e. how much the displacement of a single actuator changes the observed wavefront.

4. Conclusions

Thermal Compensation System is fundamental to contrast optical aberration arising from thermal effects when the input power in the detector increases. Its sensing system is able to identify and quantify the amount of aberrations, while their actuators can act in different and versatile ways to minimize the noise arising from thermal effects.
Looking at the next generation of the detectors, in order to further improve the efficiency of the actuation, an enhancement of the present actuators is under investigation with the design of a Mode Cleaner cavity for the CO_2 lasers and the characterization of the performances of a DM to improve the present SS actuator.

References

[1] Abbott B P et al. 2016 *Phys. Rev. Lett.* **116** 6
[2] Abbott B P et al. 2016 *Phys. Rev. Lett.* **116** 24
[3] Abbott B P et al. 2017 *Phys. Rev. Lett.* **118** 22
[4] Abbott B P et al. 2017 *Astrophys. J. Lett.* **851** 2
[5] Abbott B P et al. 2017 *Phys. Rev. Lett.* **119** 14
[6] Abbott B P et al. 2017 *Phys. Rev. Lett.* **119** 16
[7] Abbott B P et al. 2017 *Astrophys. J. Lett.* **848** 2
[8] Rocchi A et al. 2012 *J. Phys. Conf. Ser.* **363** 01
[9] Virgo collaboration 2012 Advanced Virgo technical design report *Virgo Technical Document System* VIR-0128A-12
[10] Brooks A. 2007 Hartmann wavefront sensors for advanced gravitational wave interferometers (PhD Thesis)
[11] Lück H et al. 2004 *Class. Quantum Grav.* **21** 5
[12] Ramette J et al. 2016 *Appl. Opt.* **55** 10
[13] Accadia T et al. 2011 *Proc. 12th Marcell Grossmann Meeting* ed T. Damour (Singapore: World Scientific)
[14] Rocchi A 2014 *Advanced Interferometers and the Search for Gravitational Waves* (Springer) p.251-274
[15] Lawrence R C 1997 Active Wavefront Correction in Laser Interferometric Gravitational Wave Detectors (PhD Thesis)

[16] Black E D 2001 *Am. J. Phys.* **69** 1
[17] Virgo Collaboration 2016 A Vision Beyond the Advanced Virgo Project *Virgo Technical Document System* VIR-0136A-16
[18] www.adaptica.com

IOP Conf. Series: Journal of Physics: Conf. Series **1226** (2019) 012020 doi:10.1088/1742-6596/1226/1/012020

Monte Carlo simulation of the LIDAL-ALTEA detector system

C Berucci[1], C De Donato[1,2], L Di Fino[1], C La Tessa[3,4], G Masciantonio[1], R Messi[1,2], M C Morone[1,2], A Rizzo[1,2], M Rovituso[3], F Tommasino[3,4] and L Narici[1,2]

[1] University of Rome Tor Vergata, Physics Department, Via della Ricerca Scientifica 1, 00100 Rome, Italy

[2] INFN Sezione Roma Tor Vergata, Via della Ricerca Scientifica 1, 00100 Rome, Italy

[3] Trento Institute for Fundamental Physics and Application - TIFPA INFN - Via Sommarive 14, 38123 Povo (Trento), Italy

[4] University of Trento, Physics Department, Via Sommarive 14, 38123 Povo (Trento), Italy

E-mail: carolina.berucci@roma2.infn.it

Abstract. The LIDAL (Light Ion Detector for ALTEA) is a device designed to work paired with three silicon detector units of ALTEA (Anomalous Long Term Effects on Astronauts) in order to improve the particle identification capabilities of ALTEA on the International Space Station also providing Time-of-Flight measurements. The LIDAL-ALTEA goal is to measure ions from protons up to iron in real time. The improved measurements of the radiation environment inside ISS will be very valuable for radiation risk assessment and mitigation. It is necessary to have a detailed simulation of the apparatus response to cosmic ray nuclei in order to assess the detector response, its observational capabilities and to set the relevant parameters of the device. Here a new Monte Carlo simulation of the LIDAL-ALTEA setup and physics processes, in the framework of FLUKA, is presented. A comparison between Monte Carlo simulations and calibration data is also shown.

1. Introduction

The ALTEA (Anomalous Long Term Effects on Astronauts) experiment, developed by the Italian Space Agency (ASI), was aimed at the study of the transient and long term effects of cosmic particles on the astronauts cerebral functions [1]. ALTEA [2] has measured the particle flux in the U.S. Laboratory on the International Space Station (ISS) from 2006 to 2012, being able to discriminate the type of particles, to measure their trajectories and the delivered energies.

The LIDAL-ALTEA apparatus is an upgrade of the ALTEA detector scheduled to fly in the International Space Station (ISS) in 2019; its main goal is to measure in detail also the low-Z part of the particles spectrum inside the ISS, through Time-Of-Flight (TOF) technique, measuring the velocity β of the passing through particles.

Another important task of LIDAL-ALTEA is the Particle IDentification (PID). Indeed, matching β measured by LIDAL detector with the deposited energy (dE/dx) measured by ALTEA for each event, it will be possible to efficiently identify the different elemental species providing further help for a reliable risk assessment for astronauts in space. This paper describes

IOP Conf. Series: Journal of Physics: Conf. Series **1226** (2019) 012020 doi:10.1088/1742-6596/1226/1/012020

the MC techniques used as well as the results obtained for the two configurations, ALTEA and LIDAL-ALTEA, where the first one is used as a validation for the simulation code.

2. LIDAL-ALTEA setup

LIDAL detector will consist of three modules: two LDU (LIDAL Detector Unit) placed at a distance of 49 cm, with 8×(80×20×4 mm) scintillators each and one LCU (LIDAL Collector Unit), for data acquisition. LIDAL first prototype has been developed using plastic scintillator EJ230 (Eljen Technologies [3]) for fast timing application coupled with two PMMA light guides and two HAMAMATSU PMTs RS9880U210 [4]. The plastic scintillators are characterized by a rise time of 500 ps and a good light output of 9700 photons for 1 MeV electron. The PMTs, with a rise time of 520 ps and high quantum efficiency, assure the required fast response to perform TOF measurements with the needed detail. The generated signal is then processed by Front-End Electronics, in particular by the NINO chip [5]. According to the LIDAL prototipe tests at Trento Institute for Fundamentals Physics Applications (TIFPA) proton beam line, we expect to measure the particle TOF with a resolution better than σ = 120 ps [6] and to provide trigger for the light and fast particles to the whole detector system. In between the two LDU modules, three Silicon Detector Unit (SDU) of the ALTEA system will be positioned (see Fig. 1). The ALTEA detector system consists of a set of six SDUs, silicon strip particle telescope able to measure, in real time, the energy released by the ions and the trajectory of each incoming particle within the triggering window. Every SDU is composed of six planes, alternatively stripped along the X and Y axis with an active area of 2×(8×8) cm^2 and a thickness of 380 μm. Each silicon chip is segmented in 32 strips with a pitch of 2.5 mm. The distance between an X and the correspondent Y plane is 3.75 mm, while the distance between each XY couple is 37.5 mm. The digital signal is driven to the Read Out Electronics (ROE) board, placed as a 7th plane of the telescope. The modularity of ALTEA detector system permits to work with different configurations, depending on the experiment involved. LIDAL-ALTEA system will use only three SDUs, a sketch of the experimental setup is shown in Fig. 1.

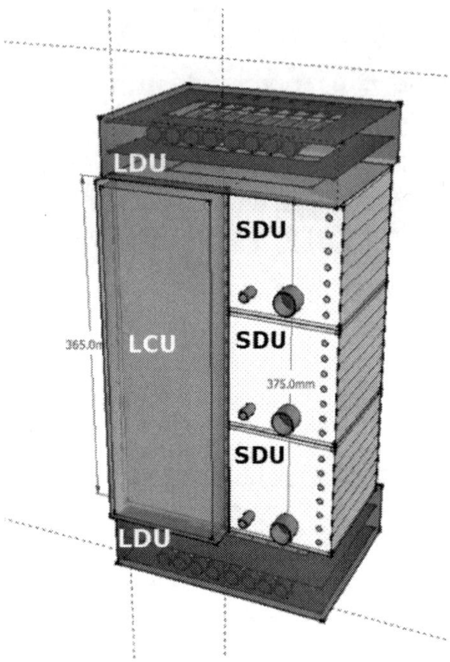

Figure 1. LIDAL-ALTEA setup. LIDAL detector consists of three modules: two LDU (LIDAL Detector Unit) and one LCU (LIDAL Collector Unit). In between the two LDU modules, three Silicon Detector Unit (SDU) of the ALTEA system are located.

3. ALTEA test-beam preliminary results and MC simulation

A detailed Monte Carlo simulation of the detector system response to cosmic ray particles is necessary in order to evaluate its observational capabilities. The LIDAL-ALTEA apparatus was therefore simulated using a new FLUKA [7, 8] code. To achieve the best description of the data, the parameters of simulations are tuned with different methods. The simulation procedure started with the setup configuration already used during the ALTEA calibration test at the proton beam line at TIFPA in June 2016. Three different proton beams of 70, 91 and 119 MeV have been used to verify, after the ALTEA reentry, the performance of the detectors (SDU1,

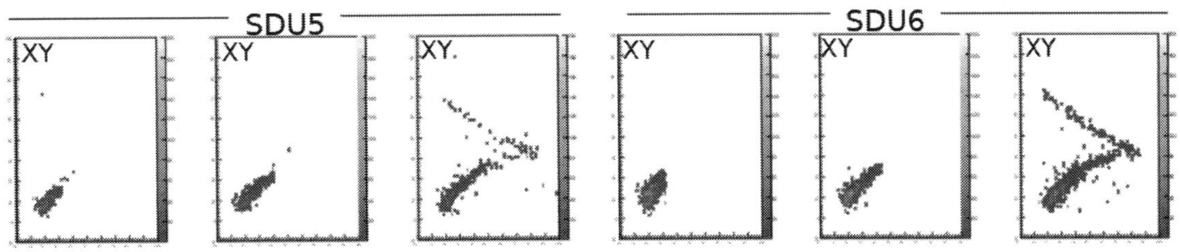

(a) data

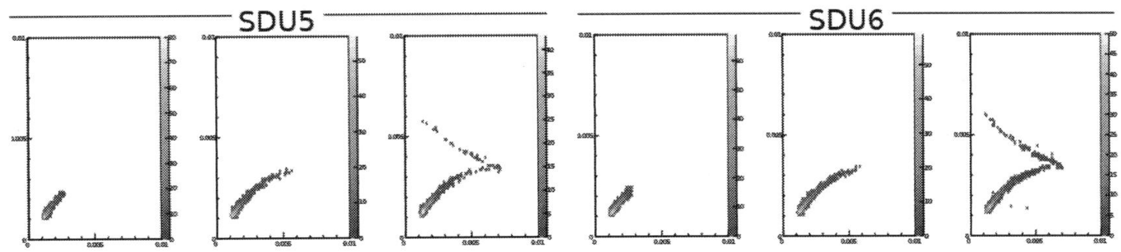

(b) Monte Carlo simulation

Figure 2. Energy release by 119 MeV protons, measured (top) and simulated (bottom), in twelve consecutive ALTEA planes corresponding to SDU5 and SDU6.

SDU2, SDU3, SDU4, SDU5 and SDU6) and the full electronics readout chain. A complete scan of each SDU was performed in order to check the functionality of all strips of the silicon detector telescope. All the beam settings have been simulated with FLUKA and the released energies in each SDU planes have been compared with the test beam data results. This procedure allows a strong check of the Monte Carlo calculation and at the same time, provides an important method to understand the critical points of ALTEA data analysis.

The structure and the composition of different material out of which the setup is built were reproduced with accuracy. The input parameters of the simulation, as geometrical and beam momentum spread, have been adjusted in such a way as to produce ΔE-E correlation with the same population of energies as in the test data. The ΔE-E technique is based on the measurement of the energies deposited in two detectors and is applicable when particles with kinetic energy E pass through the first detector and are stopped in the second one. An example of deposited energy correlation between calibration data and Monte Carlo simulation for consecutive XY silicon planes of SDU5 and SDU6 is shown in Fig. 2. In all spectrum, there are a large fraction of protons which have enough energy to pass through both silicon layers and reach the next SDU. They are responsible for a well-known characteristic "punch-through" line [9]. The III and VI plots correspond respectively to the deposited energy in the last XY silicon layers of SDU5 and SDU6; in this case, besides the passing protons, we can distinguish a near-diagonal lines. This is attributed to the protons which have an energy low enough to get stopped into the last 380 μm of silicon, while the reason why we can only measure stopped proton in the last SDU plane, depends on the ALTEA trigger logic.

4. LIDAL-ALTEA trigger system

ALTEA is triggered by incoming particles that cross the detector and release at least 1.09 MeV in all X-planes. Each SDU has an auto-trigger, with the signal compared with a ROE threshold and working in AND logic between the signal from the X-planes. LIDAL (LDU1, LDU2) is also triggered by passing through particles. Monte Carlo simulations with different proton energies

IOP Conf. Series: Journal of Physics: Conf. Series **1226** (2019) 012020 doi:10.1088/1742-6596/1226/1/012020

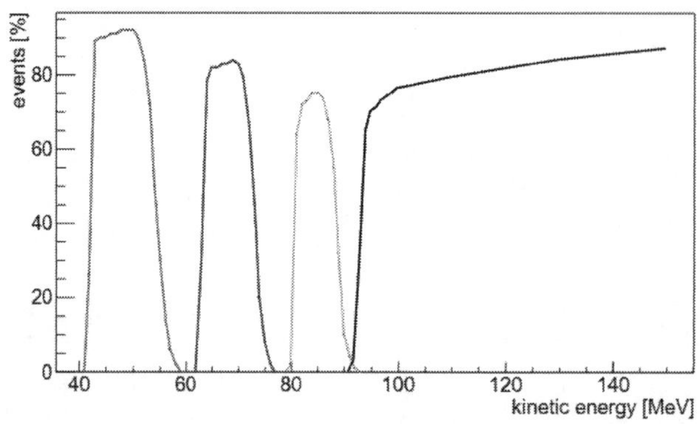

Figure 3. Simulated trigger thresholds for each of the LIDAL-ALTEA system component: (left to right) SDU1, SDU2, SDU3 and LIDAL.

have been performed in order to assess energy losses by particles and identify the energy intervals detectable by the LIDAL-ALTEA trigger system. The detector system features an energy range defining regions of incoming proton energy triggering one or more of the four components (SDU1, SDU2, SDU3 and LIDAL). Fig. 3 describes this behavior. Detection range for each proton was obtained considering the maximum incidence energy, which corresponds to the lowest energy released by the particle (trigger threshold), and the minimum incidence energy, corresponding to the maximum energy loss. In particular, the minimum energy value is, for light particles like proton, the minimum energy that is necessary to reach the last plane of the detector and trigger the device while, for heavier particles, the energy that drives the device to saturation. One can notice that ALTEA SDUs are able to trigger protons with energy between 43 and 91 MeV, while LIDAL cover the rest of the protons spectrum.

5. Conclusion

In this preliminary work a comparison between calibration data results acquired with six SDUs and Monte Carlo simulations have been reported. Test results are in agreement with the device expected performances and have been used for energy calibration and for tuning the MC simulation. Finally, we can say that dE/dx and consequently the punch-through reconstruction was verified by MC simulations. These simulations, validated with the ALTEA calibration data, will be used to prepare the next beam test at TIFPA and GSI facilities needed to fully describe the new features of the integrated LIDAL-ALTEA detector. A following step will include a throughout simulation of radiation environment in ISS and an assessment of LIDAL-ALTEA behavior in orbit.

6. References

[1] Narici L *et al* (2008) *New Journal of Physics* **10** 075010
[2] Zaconte V *et al* (2010) *Rad. Meas.* **45** 168?72
[3] https://eljentechnology.com/products/plastic-scintillators/ej-228-ej-230
[4] https://www.hamamatsu.com/resources/pdf/etd/R9880UT-PMH1321E.pdf
[5] Aghinolfi F *et al* (2004) *IEEE Trans. Nucl. Sci.* **51** 1974-1978
[6] Rizzo A *et al* (2018) *Nucl. Instrum. Meth. A***898** 98-104
[7] Ferrari A *et al* (2005) CERN-2005-010; INFN-TC-2005-11;SLAC-R-77
[8] Battistoni G *et al* (2007) *AIP Conf. P roc.* **896** 31-49
[9] Carboni S *et al* (2012) *Nucl. Instrum. Meth. A* **664** 251-263

Neutron production with particle accelerators

D De Luca[1]*, L Campajola[1,2], P Casolaro[1,2]

[1]Department of Physics "E. Pancini", University of Naples Federico II, Naples, Italy

[2]National Institute for Nuclear Physics (INFN), Naples Section, Italy

*Email: delucadaniela9@gmail.com

Abstract. Neutron beams have been produced at Laboratory of High Energy Physics (LHEP) of University of Bern. LEHP is equipped with an 18 MeV cyclotron with an external beam line. Simulations, analytical calculations and kinematics studies have been carried out, in order to produce several neutron beam configurations with characteristics suited for specific applications. Furthermore, a procedure for energy modulation has been developed to match the fixed cyclotron energy to the needed requirements. The technique described in this work is of general application and can be applied to more intense beams up to 10^{10} pps.

1. Introduction

Neutrons are of primary interest in many areas of basic and applied physics. Their production and detection still represents an uneasy task. This is due to its peculiar behavior, which is very different from that of charged particles, X and gamma rays. Neutrons interact with matter through nuclear force only and the reaction cross sections are strongly energy dependent [1].

Neutrons can be produced by means of radioactive sources (such as ^{252}Cf), nuclear reactors or particle accelerators. The major disadvantage in using nuclear reactors lies in the huge amount of gamma rays produced both by fission and interactions between neutrons and fission fragments in the core. This results in a very broad spectrum, which is not always desired. Instead, different types of incident particles can be used with particle accelerators to produce neutrons. These are, for example, protons, alpha particles, deuterium or tritium. In this work, the neutron production was carried out with protons from a cyclotron.

One of the main reasons why different aspects of neutrons production have been studied in this work is related to the low number of facilities dedicated to this topic. In fact, in Italy there are no facilities specifically qualified for the neutron production. Nevertheless, Italy is the third user in the world of ISIS (Neutron and Muon Source - Oxford UK) and the fifth European nation in number of publications [2].

A second motivation for this work is the growing interest in nuclear applications involving neutrons in a very wide energy spectrum (from thermal to fast and relativistic neutrons). Neutron applications cover several fields, ranging from industry to research and medicine: fast neutron cancer therapy, isotope production, measurement of dose administrated to aviation personnel, heritage studies on precious and delicate objects or works of art, radiation hardness, science and chemistry of materials, material recognition, nuclear astrophysics and so on. A considerable requirement comes from the field of basic research: it is an example the SPES project at Legnaro National Laboratories (LNL) [3].

New production methods have been studied in order to carry out neutron beams with the best characteristics suitable for the users.

2. Preparatory study

The characterization of neutron beams produced with accelerators needs measurements of the energy spectrum at different angles. The energy of the neutrons produced is strictly related both to the features of the target (density, thickness, etc ...) and the energy of the projectiles. Since the production rate can vary a lot with the solid angle covered by the detector, it is fundamental to know the distance between the source and the detector.

Several targets have been used to produce the neutron beams. We identified the appropriate energy range for the targets studied by calculating the output energy of the neutrons. The simulations needed for the calculation of the reactions cross section, kinematics and energy loss in the media were made with *cross section plotter*, *Trim* and *Lise++* [4, 5]. On the basis of the simulation results, the materials analyzed as potential target for neutron production are the following: ^{7}Li, ^{27}Al, ^{181}Ta, nat-Ag, ^{197}Au, ^{93}Nb. These are characterized by a limited and low energy range useful for the occurrence of the (p, n) reaction only. Since the cyclotron produces 18 MeV protons with fixed energy, in order to have less energetic beams, the incident energy must be appropriately degraded. For this reason, it is necessary to find a material that allows the degradation of the beam energy without producing unwanted neutrons. The best material has been found to be the graphite. In fact, unlike other materials which are characterized by several neutron producing reactions at the proton energy (18 MeV) it presents only the reaction (p, n + ^{3}He). The only problem in using ^{12}C is related to the high cross section (~ 6 barns) of the above mentioned reaction at the beam energy (Fig. 1). This implies that each acquired spectrum has a neutron background from the graphite that must be subtracted. Because of the three-body nature of this reaction, the only way to evaluate the contribution of graphite comes from simulations.

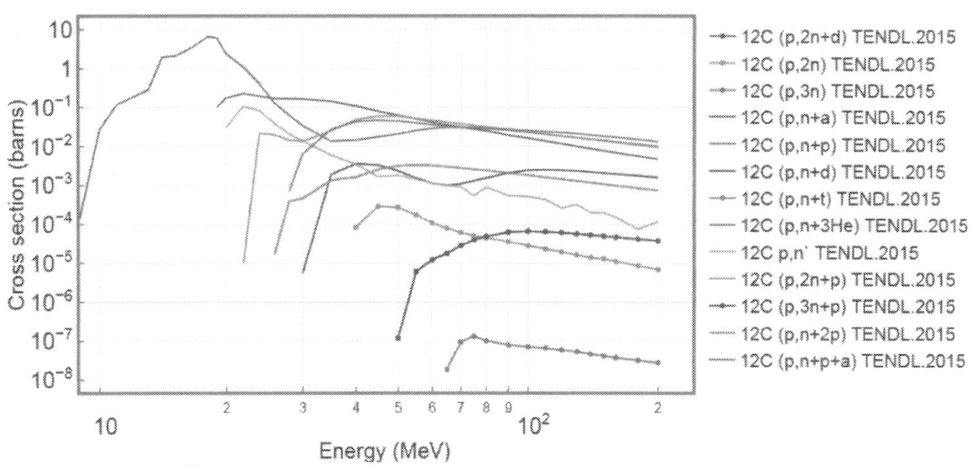

Figure 1: Cross section values as a function of the energy for protons in graphite [6].

3. Simulation results

In addition to the previous simulations, the design of the experiment was accomplished with the Monte Carlo simulation package FLUKA [7]. Simulations were carried out at first to evaluate the effectiveness of graphite.

As an example, in the Fig. 2a we show the neutron spectrum of an 18 MeV proton beam on a graphite target, and we compare it with the one in which a thin Lithium layer was added (Fig. 2b). As is evident, the presence of the Lithium layer, although very thin (50 μm), changes the shape of the spectrum: it behaves like a neutron moderator. The 4.5 MeV peak in fig. 2b refers to the neutrons produced in Lithium.

The same process is done with the other targets; then other elements were added, up to reach more complex configurations that better describe the geometry of our experiment.

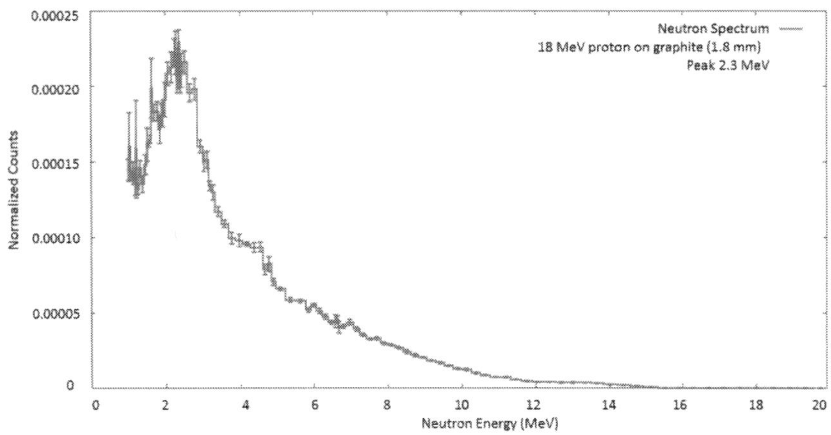

a)

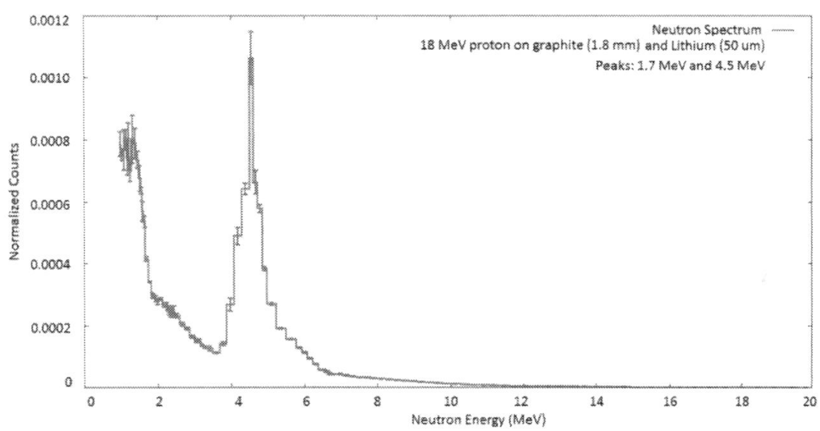

b)

Figure 2: Results of Fluka simulation for an 18 MeV proton beam on graphite layer (a) and a combination of graphite and Lithium (b).

4. The experiment

The neutron production was carried out for the first time at the IBA 18/9 Cyclone of the University of Bern LHEP. The 18/9 cyclotron is mainly used for the production of radioisotopes for nuclear medicine. It is equipped with a Beam Transport Line (BTL) on which there are many devices useful for the beam quality and focusing controls. A schematic of the main elements of the BTL is reported in Fig. 3.

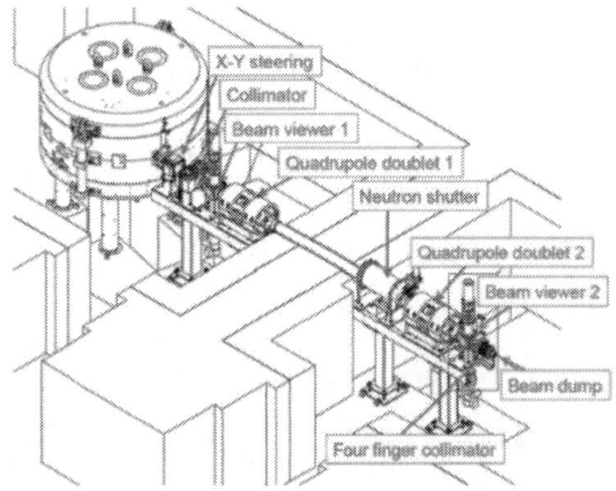

Figure 3: Schematic view of the Bern Cyclotron facility [8].

5. The target design

The target has been designed as shown in Fig. 4a): graphite degrader, target for neutron production and beam stopper (graphite). The second layer of graphite does not represent a problem in terms of neutron production, in fact at low energies there are no neutrons produced (Fig. 4b).

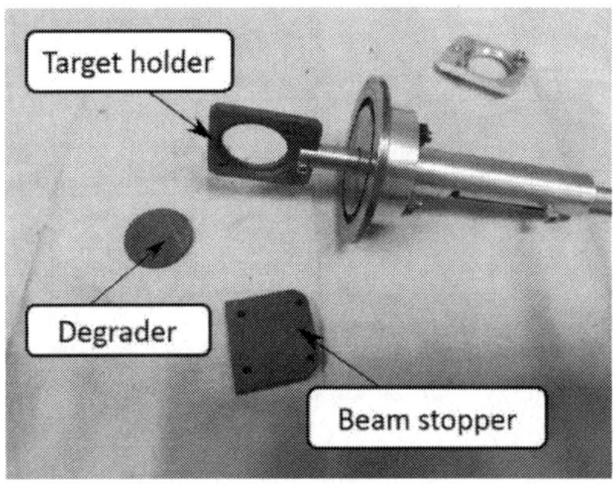

a)

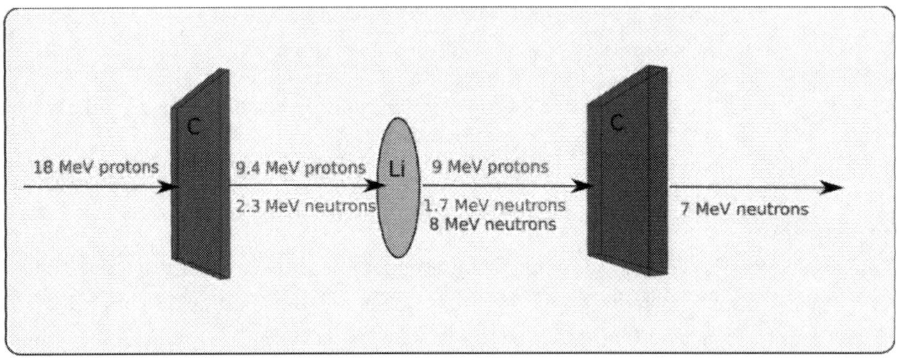

b)

Figure 4: (a) Target designed for the experiment: target holder, beam stopper and degrader. (b) Scheme of the several processes which happen in each step of the multiple target.

6. The production

Measurements were made to produce neutron beams of different energies from different targets.

The detector used is the Berthold neutron dose-rate meter (Berthold LB 6411 probe) [9]. It is useful for measurements of the equivalent ambient dose for neutrons and consists of a moderating polyethylene sphere with a composite ^{3}He recoil proton counter tube, the LB 6410, at its center. The instrument has an extremely high sensitivity of approximately 3 counts/nSv, while the response for gamma radiation is approximately 10^3 counts/nSv.

The targets are positioned in the beam line as described above. The Berthold dosimeter is 80 cm far from the target and it is positioned at 0 degrees from the beam axis. The results of the measurements made with all the targets are compared with the analytical calculation (table 1) and shown in Fig. 5.

The intercept value different from zero corresponds to the neutrons produced in graphite. This is a background and a constant offset that has to be subtracted for all the measurements.

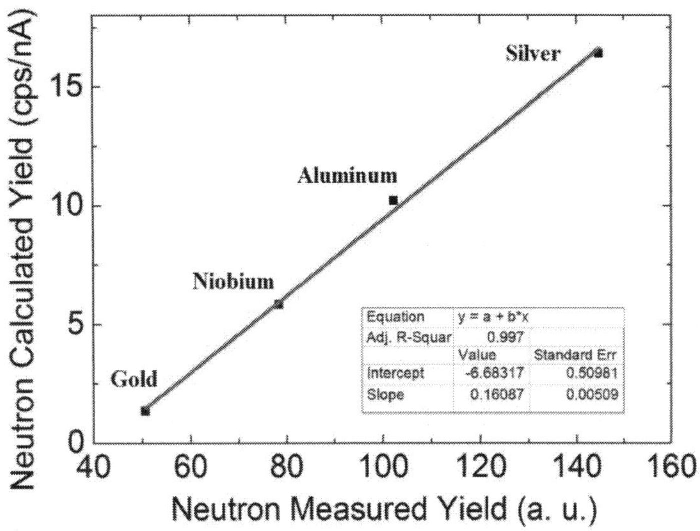

Figure 5: Linear fit (red line) between the calculated and the measured neutron yield for: Gold, Niobium, Aluminum and Silver. Data from the Table 1.

Target	Measured cps/nA	Calculated cps/nA
Nb	78.49 ± 1.16	5.83
Au	50.63 ± 0.80	1.35
Ag	144.74 ± 1.38	16.40
Al	102.30 ± 1.14	10.2

Table 1: Measured and calculated counts per second, normalized for the beam current (nA).

7. Conclusions

The measured neutron yields, normalized to the beam current, are in agreement with the analytical calculations, taking into account the contribution of neutrons produced in graphite.

This work opens interesting perspectives in the study of new methods for neutrons production and of their applications. First of all, the method investigated in this work, can be applied with intense proton beams, in order to reach neutron production rate up to 10^{10} pps. Improvements on the method could be obtained by optimizing the target design in order to use other kinds of target, including water cooled targets for high intensity beams.

Regarding the applications, this technique can be used in: cross section measurements for neutron production in different materials; setting up of a facility for testing of new generation detectors to be used in applied and fundamental Physics; use of neutron beams to produce radioisotopes in Medical Physics (^{99}Tc); design of a dedicated tunable neutron facility; use of high intensity proton beams for applications in many fields (Radiation Hardness).

References

[1] Knoll G F *Radiation Detection and Measurements* (John Wiley & Sons)
[2] *https://www.isis.stfc.ac.uk/Pages/home.aspx*
[3] De Angelis G, Prete G, Andrighetto A, Manzolaro M, Corradetti S, Scarpa D and Vasquez J The SPES project of INFN: Facility and detectors, **88**, *EPJ Web of Conferences*
[4] Ziegler J F, Ziegler M D and Biersack J P SRIM – The stopping and range of ions in matter *NIM Section B* **268**, 1112
[5] Tarasov O B and D Bazin 2003 Development of the program LISE: application to fusion– evaporation *NIM Section B,* **204**, 174178
[6] Shimwell J *Cross section plotter* http://www.cross-section-plotter.com/
[7] Ferrari A *et al.* 2005 FLUKA: A multiparticle transport code *INFN-TC-05-11*
[8] Nesteruk K P, Auger M, Braccini S, Carzaniga T S, Ereditato A and Scampoli P 2018 A system for online beam emittance measurements and proton beam characterization *JINST 13 P01011.*
[9] Burgkhardt B, Fieg G, Klett A, Plewnia A and Siebert B The Neutron Fluence and H*(10) response of the New LB 6411 REM Counter *Radiation Protection Dosimetry*, **70**, 1-4

IOP Conf. Series: Journal of Physics: Conf. Series **1226** (2019) 012022 doi:10.1088/1742-6596/1226/1/012022

Final results of the OPERA experiment on ν_τ appearance and the OPERA "legacy"

Giuliana Galati[1]

[1]INFN Sezione di Napoli, I-80126 Napoli, Italy

E-mail: giuliana.galati@na.infn.it

Abstract. OPERA (Oscillation Project with Emulsion tRacking Apparatus) was a long-baseline experiment at the INFN Gran Sasso laboratory (LNGS) designed to search for $\nu_\mu \to \nu_\tau$ oscillations in appearance mode. It took data from 2008 to 2012 with the CNGS neutrino beam from CERN. In 2015, after the detection of five ν_τ candidates with a signal-to-background ratio of $\sim$ 10, the discovery of ν_τ appearance in the CNGS beam was announced with 5.1σ significance. After having reached the experiment main goal, the selection of ν_τ candidates has been extended by loosening the selection criteria and applying a multivariate approach for events identification, in order to improve the statistical uncertainty in the measurement of the oscillation parameters and of ν_τ properties. Future experiments that will take advantage of the improvements done by OPERA in the use of nuclear emulsions will also be described.

1. Introduction

Originated by the neutrino mass and the mixing between flavour and mass eigenstates, neutrino oscillations are now established thanks to intense experimental efforts. In 1998, the first evidence for neutrino oscillations was provided by the Super-Kamiokande experiment, showing the disappearance of muon neutrinos produced in the atmosphere [1]. This result was interpreted as a possible transition from ν_μ to ν_τ or to a new type of neutrino, still not known. At that time, moreover, the ν_τ neutrino had not been observed yet. Therefore, OPERA (Oscillation Project with Emulsion tRacking Apparatus) was conceived to demonstrate unambiguously the transition from ν_μ to ν_τ by directly observing the τ lepton produced in charged-current (CC)

IOP Conf. Series: Journal of Physics: Conf. Series **1226** (2019) 012022 doi:10.1088/1742-6596/1226/1/012022

ν_τ interactions in a pure ν_μ beam [2]. It was operated underground in the Gran Sasso INFN Laboratory (LNGS), 730 km away from the muon neutrino source.

The challenge of the OPERA experiment to detect the short-lived τ lepton ($c\tau =$ 87 μm), produced in the CC ν_τ interactions, was achieved using the nuclear emulsion technique that features micrometric spatial resolution.

2. The CNGS beam and the OPERA detector

Runs with the CNGS neutrino beam [3, 4] were successfully carried out from 2008 to 2012, with a total beam intensity of $17.97 \cdot 10^{19}$ protons on target (p.o.t.). The average neutrino energy was about 17 GeV, the contamination from $\bar{\nu}_\mu$ interactions was 2.1%, the one from ν_e and $\bar{\nu}_e$ together below 1%, while the number of prompt ν_τ negligible. The detector was an hybrid apparatus consisting of an emulsion/lead target complemented by electronic detectors. It was made up of two identical super-modules aligned along the CNGS beam direction, each made of a target section and a muon spectrometer. Each target section consisted of a multi-layer array of 31 target walls interleaved with pairs of planes of plastic scintillator strips. Target walls were made of Emulsion Cloud Chamber target units, called bricks, which were, in total, 150000. Each brick consists of 57 emulsion films, 300 μm thick, interleaved with 56 lead plates, 1 mm thick, for a total mass of 8.3 kg. The electronic detectors were used to identify the brick containing the neutrino interaction, for muon identification and its charge and momentum determination.

The appearance of the τ lepton was identified by the detection, in the nuclear emulsions, of its characteristic decay topologies, either in one prong (electron, muon or hadron) or in three prongs. Kinematic selection criteria were then applied according to the decay channel.

3. First phase of the OPERA experiment

In 2015, five ν_τ candidates [5–9] had been observed, satisfying stringent kinematical selection criteria as reported in [6]. The observation of these five ν_τ candidates with a background of 0.25 ± 0.05 events, allowed to exclude the absence of $\nu_\mu \rightarrow \nu_\tau$ oscillations with a 5.1σ significance, as reported in the paper [9].

4. Second phase of the OPERA experiment

After discovering the ν_τ appearance in the ν_μ CNGS beam [9], a new goal has been set: estimate the oscillation parameters in appearance mode and exploit the unique capability of the OPERA experiment to identify all three neutrino flavours in order to put constrains on the oscillation parameters by a joint oscillation fit of all datasets.

Given the validation of the Monte Carlo simulation of ν_τ events, based on different control data samples, a new analysis strategy was developed, fully exploiting the features expected for ν_τ events. A multivariate approach to improve signal to noise separation was applied to candidate events selected by means of moderately tight topological and kinematical cuts.

Details about the new selection method are reported in [10]. The total expected signal, after selection criteria application, is (6.8 ± 1.4) events, whereas the total background expectation is (2.0 ± 0.4) events. Ten events have been observed [10].

Different multivariate techniques have been considered and their performances for signal to background discrimination compared. The one with the best discrimination power was the Boosted Decision Tree (BDT).

4.1. Results

The statistical analysis used to re-evaluate the significance for the ν_τ appearance is based on an extended likelihood constructed as the product of a probability density function given by the BDT response, a Poisson probability term which takes into account the number of observed events and the expected background in each decay channel, and a Gaussian term which accounts for systematics. The discovery significance of ν_τ appearance is expressed in terms of a hypothesis test where the background only hypothesis plays the role of the null hypothesis and the signal-plus-background hypothesis is the alternative one. The null hypothesis is now excluded with the improved significance of 6.1 σ [10].

The number of observed ν_τ candidates after background subtraction is a function of the product of ν_τ CC cross-section ($\sigma^{CC}_{\nu_\tau}$) and the oscillation parameter Δm^2_{23}. Δm^2_{23} has been evaluated for the first time in appeareance mode: assuming the other oscillation parameter $\sin^2 2\theta_{23} = 1$, $|\Delta m^2_{32}|$ is equal to $(2.7^{+0.7}_{-0.6}) \cdot 10^{-3}\text{eV}^2$. The result is consistent within 1σ with the measurements performed in disappearance mode by other experiments and with the Particle Data Group best fit [11].

ν_τ CC cross-section on the lead target has been estimated to be equal to $(5.1^{+2.4}_{-2.0}) \cdot 10^{-36}\text{cm}^2$, when $|\Delta m^2_{32}| = 2.50 \cdot 10^{-3}$ eV^2. This result is the first measurement of the ν_τ CC cross-section with a negligible contamination from $\bar{\nu}_\tau$.

The lepton number of ν_τ has never been observed. In the muonic channel, the OPERA experiment can distinguish neutrinos from anti-neutrinos looking at the charge of the muon produced in τ decays, which was measured to be negative at 5.6 σ level for the $\tau \to \mu$ candidate. Performing a dedicated BDT analysis which included also the background due to the 2% $\bar{\nu}_\mu$ beam contamination, we obtained the first direct evidence for the leptonic number of τ neutrinos with a significance of 3.7σ. All these results have been reported in [10]

5. The OPERA legacy

The improvements achieved for the OPERA experiment in the use of nuclear emulsions and in the scanning systems to analyse them make this kind of detector still attractive in a wide range of scientific fields and applications. Hereafter some of the future experiment will be briefly outlined.

5.1. The SHiP experiment at CERN

The Standard Model provides an explanation for many subatomic processes, but fails to explain some observed phenomena such as the so-called dark matter, the neutrino oscillation and masses and the matter/antimatter asymmetry in the Universe. The SHiP experiment, a multi-pourpose facility proposed at CERN, is aimed at searching for very weakly interacting long lived particles including Heavy neutral leptons. Moreover, the facility is also suited to study ν_τ and $\bar{\nu}_\tau$ interactions with a nuclear emulsion detector [12].

5.2. The Nuclear Emulsions for WIMP Search with directional measurement (NEWSdm)

The NEWSdm experiment aims at the direct detection of WIMP dark matter candidates by measuring the direction of WIMP-induced nuclear recoils in the nuclear emulsions. For this challenge, the detector exploits new generation nuclear emulsions with nanometric grains and microscopes that can overcome the diffraction limit, reaching a resolution of 10 nm [13].

5.3. The FOOT (FragmentatiOn Of Target) experiment for hadron therapy

Particle therapy uses proton or C^{12} beams for the treatment of deep-seated solid tumors. Nuclear fragmentation of both beam and patient nuclei must be carefully taken into account, but this process is not known with enough precision at present. The FOOT experiment, thanks also to the nuclear emulsions in its detector, will provide an accurate description of nuclear interactions taking place between the beam and the patient tissues during the treatment [14].

5.4. Muon Radiography

Muon Radiography is a technique that uses cosmic ray muons to generate three-dimensional images of volumes using information contained in the Coulomb scattering of muons. Nuclear emulsions are a very portable detector, since they do not need electric supply, therefore they are used to obtain radiography of volcanoes [15], pyramids [16] and other archeological sites of interest.

References

[1] Kajita T (Super-Kamiokande) 1998 *New era in neutrino physics. Proceedings, Satellite Symposium after Neutrino'98, Tokyo, Japan, June 11-12, 1998* 107–122
[2] Acquafredda R *et al.* 2009 *JINST* **4** P04018
[3] Acquistapace G *et al.* 1998 *CERN-98-02, INFN/AE-98/05*
[4] Baldy R *et al.* 1999 *CERN-SL-99-034-DI, CERN-SL-99-34-DI, INFN-AE-99-05, INFN-AE-99-5*
[5] Agafonova N *et al.* (OPERA Collaboration) 2010 *Phys.Lett.* **B691** 138–145 (*Preprint* 1006.1623)
[6] Agafonova N *et al.* (OPERA Collaboration) 2013 *JHEP* **11** 036 [Erratum: JHEP **04**, 014(2014)] (*Preprint* 1308.2553)
[7] Agafonova N *et al.* (OPERA Collaboration) 2014 *Phys. Rev.* **D89** 051102 (*Preprint* 1401.2079)
[8] Agafonova N *et al.* (OPERA Collaboration) 2014 *PTEP* **2014** 101C01 (*Preprint* 1407.3513)
[9] Agafonova N *et al.* (OPERA Collaboration) 2015 *Phys. Rev. Lett.* **115** 121802 (*Preprint* 1507.01417)
[10] Agafonova N *et al.* (OPERA) 2018 *Phys. Rev. Lett.* **120** 211801 (*Preprint* 1804.04912)
[11] Patrignani C *et al.* (Particle Data Group) 2016 *Chin. Phys.* **C40** 100001
[12] Anelli M *et al.* (SHiP) 2015 (*Preprint* 1504.04956)
[13] Aleksandrov A *et al.* (NEWS) 2016 (*Preprint* 1604.04199)
[14] Battistoni G *et al.* 2017 *PoS* **BORMIO2017** 023
[15] Tioukov V *et al.* 2017 *Annals of Geophysics* **60** S0111
[16] Morishima K *et al.* 2017 *Nature* **552** 386–390 (*Preprint* 1711.01576)

IOP Conf. Series: Journal of Physics: Conf. Series **1226** (2019) 012023 doi:10.1088/1742-6596/1226/1/012023

Thermal noise reduction for future gravitational wave detectors

Diana Lumaca[1,2] on behalf of Virgo Coating R&D Collaboration

[1] Università di Roma Tor Vergata, Via della Ricerca scientifica 1, I-00133 Roma, Italy
[2] INFN - Roma Tor Vergata, Via della Ricerca scientifica 1, I-00133 Roma, Italy

E-mail: diana.lumaca@roma2.infn.it

Abstract. The coating thermal noise limits the sensitivity in gravitational wave detectors from few tens to hundreds Hz, where first gravitational signals were detected and others are expected. In view of future upgrades, the increase in the mechanical performances of reflective coatings, retaining their outstanding optical and morphological properties, is fundamental. In the Virgo collaboration a coating R&D group working with this aim is born. One of the research lines regards the mechanical characterization of both substrates and coatings, looking for materials with low mechanical loss angle. To perform a precise coating mechanical characterization, the substrate on which it is deposited must be characterized as well and must be stable with respect to its dissipative behaviour. Commercial SiO_2 substrates are subject to effects that change their mechanical condition during time, compromising the characterization. The source of these spurious losses is related to the rough lateral surface: after its polishing, this behaviour is largely reduced. We designed and assemble a facility for the CO_2 laser polishing of the substrates barrel, to provide a reliable heat treatment, reducing spurious losses to a negligible level. Other further treatments have been tested, with the aim of taking under control ϕ deterioration. We develop a procedure to prepare SiO_2 samples for coating deposition. Procedure steps, mechanical characterizations and first results are shown.

1. Introduction

On September 14th 2015 a Gravitational Wave (GW) coming from binary black holes coalescence was detected for the first time [1], a hundred years after their prediction by Einstein. Five other similar signals and a signal from binary neutron stars merger were detected, both by Advanced LIGO and Advanced Virgo intereferometers [2]. The network of the three detectors all together allowed to significantly improve distance measurement and sky localization [2, 3] and the binary neutron stars signal was also followed by detection of electromagnetic counterpart, arrived to the Earth 1.7 second after [4].
Enhance the sensitivity of interferometric detectors is needed to obtain greater numbers of detections at lower noise levels, opening the doors to GW astronomy. Both current and future interferometers, in the middle frequency band, from few tens to hundreds Hz, are limited by mirror thermal noise. The main contribution to this noise is Coating Thermal Noise (CTN), related to structural dissipation inside the multilayer dielectric coatings [5]. The reduction of this noise is fundamental for advanced detectors. More generally, finding a way to reduce CTN would be beneficial also for other precision experiments using high-finesse optical cavities, such as frequency standards for laser stabilization, quantum computing devices and opto-mechanical resonators.

2. Thermal noise in GW detectors

The first observation of thermally-driven displacement was by Brown in 1828 [6] and it took almost a century to develop a full mathematical treatment of Brownian motion by Einstein [7]. Nowadays it is known that any parameter characterising a dissipative system will exhibit spontaneous thermal fluctuations, as described by the Fluctuation-Dissipation theorem [8]. From the theory it is possible to derive that the thermal noise power spectral density S_x of the system's displacement $x(t)$ is inversely proportional to the real part of the mechanical impedance of the system:

$$S_x(f) = \frac{k_B T}{\pi^2 f^2 Re\{Z(f)\}}, \tag{1}$$

where T is the temperature, k_B is the Boltzmann constant, f is the frequency, and $Z(f)$ is the system mechanical impedance [9]. These thermal fluctuations are represented by the damping coefficient of the system. Particularly, internal damping effects (i.e. internal frictions) arise from the anelasticity of the system itself: when a stress is applied, the strain response is not instantaneous. The stress-strain phase lag is the system damping coefficient of interest, that is the mechanical loss angle, defined as $\phi(f_0) = E_{lpc}/2\pi E_{tot} = \Delta f/f_0 = 1/Q$, where E_{tot} is the total energy stored in the oscillation system, E_{lpc} is the energy lost per cycle of the oscillation, Δf is the width of the resonance peak and Q is the mechanical quality factor. The work carried out by Levin [10] apply the general description to the case of the thermal noise of the mirrors and coating for the GW detector:

$$S_{TN}(f) = \frac{2k_B T}{\sqrt{\pi^3} f} \frac{1-\sigma^2}{w Y_{sub}} \left(\phi_{sub} + \frac{2}{\sqrt{\pi}} \frac{1-2\sigma}{1-\sigma} \frac{d}{w} \phi_{coat} \right), \tag{2}$$

where w is the radius of the laser beam at the mirror surface, Y_{sub} and σ are respectively the material Young's modulus and the Poisson's ratio and d is the coating thickness. The dominant part of mirror thermal noise is the one from the high reflecting coating. To reduce the CTN, one can act on (see Eq. 2) temperature (going cryogenic), coating thickness (increasing the index of refraction), laser spot dimension (to be increased), or finally coating loss angle ϕ_{coat} (to be reduced).

2.1. Coatings state of art

GW detectors high-reflection coatings are Bragg reflectors, composed of alternate layers of ion-beam-sputtered fused silica (SiO_2) and titania doped tantala (Ti:Ta_2O_5) as low- and high-refractive-index materials [11]. The coatings are designed and deposited in the Laboratoire de Matériaux Avancés (LMA), in Lyon, and allow to lose less than 0,0001% of the light impinging on them, having very low absorption and scattering (Tab. 1). They are the most advanced optics in the world.

Table 1. Left: properies of the coating deposited in LMA [12]. Right: optical and mechanical parameter of present coating materials [13, 14].

Flatness	<0.5 nm RMS (within ∅ 150 mm)
Thickness uniformity	0.05 % (within ∅ 150 mm)
Absorption	<0.4 ppm
Scattering	<10 ppm

Coating material	Reflective index (at 1064 nm)	Loss angle
SiO_2	1.45±0.01	$(4.5\pm0.3)e^{-5}$
Ta_2O_5	2.03±0.02	$3e^{-4}$
Ti:Ta_2O_5	2.07±0.02	$(2.4\pm0.3)e^{-4}$

Nonetheless, the mechanical losses (related to the Ta_2O_5 layers) are still at an high level and need to be reduced for future upgrades of GW detectors.

IOP Conf. Series: Journal of Physics: Conf. Series **1226** (2019) 012023 doi:10.1088/1742-6596/1226/1/012023

3. The Virgo Coating R&D collaboration

In the last decade, the path toward lower noise coatings starts to get clearer, thanks to national and international research projects, together with the theoretical guidance and technological expertise growth. Inside the Virgo collaboration the need to reduce CTN has been stressed and recently a collaboration started to work hardly on it: the Virgo Coating R&D (VCR&D) collaboration. It is made up of many research groups of the Virgo collaboration (both from Italy and France) and has within it coating deposition technologies, internal friction measurements facilities, optical and structural characterization facilities. Furthermore, the members of this collaboration have developed the nowadays employed multilayer reflective coatings and have a leading role in the world research on low-noise coatings.
VCR&D group works on many subjects, especially metrology (mechanical characterization, thermo-elasticity calculation) new coating materials (new oxides, nitrates, fluoride, new mixing and nano-layered composites), deposition parameters, with the aim of developing new multilayer reflective coatings for the Advanced Virgo upgrades and new technologies for future detectors. In this work the procedure for the mechanical characterization of the coating, achieved by loss angle measurements, is shown.

4. Low-loss material mechanical characterization

As shown through eq. 2, one way to reduce CTN is lowering coating loss angle ϕ_{coat}. This can be done finding new materials or mixtures with lower mechanical losses or improving mechanical losses research in current coating materials. In both cases, a way to mechanically characterize coating loss angle is through a differential measurement that follow this formula [15]:

$$\phi_{coat} \approx \frac{Y_{sub} d_{sub}}{3 Y_{coat} d_{coat}} (\phi_{sub+coat} - \phi_{sub}), \tag{3}$$

where Y and d are Young's moduli and thickness of substrate and coating. Therefore to analytically extract the coating loss angle, is necessary to mechanically characterize a sample both before and after the coating deposition. In this perspective, it is fundamental to have a well defined characterization procedure of the substrates.

4.1. Ring-down method and Gentle Nodal Suspension

The mechanical characterization of substrates and coatings can be performed studying small disk-shaped samples, on which different coatings material can be deposited. The ϕ measurements are performed through the ring-down method, exciting the resonant mode of the sample and measuring the exponential decrease in the free oscillation amplitude:

$$\phi = \frac{1}{\pi f \tau}. \tag{4}$$

This implies that the sample has to be held by some kind of suspension or clamping, making the coupling between the two negligible. The Gentle Nodal Suspension (GeNS) system [16], is worldwide recognized as one of the most powerful tool for mechanical characterization. In GeNS, a disk shaped sample is placed in equilibrium from its centre on top of a sphere, that provides a mechanically stable support (Fig. 1).
GeNS allows to avoid the clamping of the samples, suppressing any potential additional source of damping; only one of the two faces is touched and the contact area is minimized. Furthermore coated surfaces are not damaged and all mode families with nodal lines passing through the centre of the sample can be measured.

IOP Conf. Series: Journal of Physics: Conf. Series **1226** (2019) 012023 doi:10.1088/1742-6596/1226/1/012023

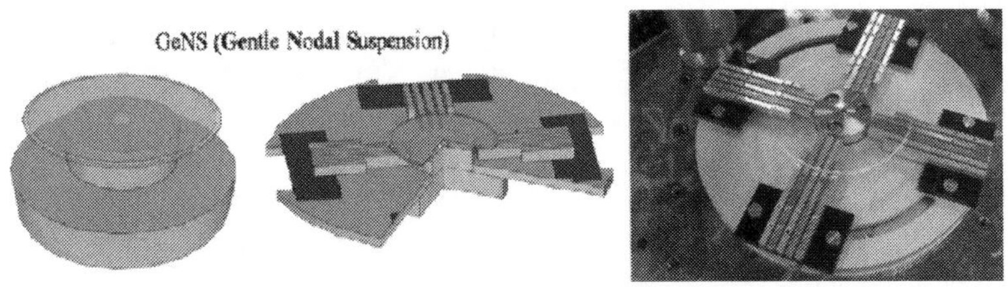

Figure 1. GeNS system design and layout.

4.2. Fused silica substrates issues

Fused silica substrates have always shown a particular behaviour in the losses distribution. It has been shown that their loss angle depends on the modes, following a distribution into families [17]; furthermore these substrates can also suffer loss angle variation over time due to ageing effects and absorption of contaminants. In Fig. 2 there is an example of these effects: in all the coloured series the separation in two branches (depending on the belonging family) is evident, on top all butterfly modes (only azimuthal nodal lines) on bottom all mixed modes (both azimuthal and radial nodal lines); following the colour sequence (blue - red - green - yellow) also the ageing/contamination effects are shown.

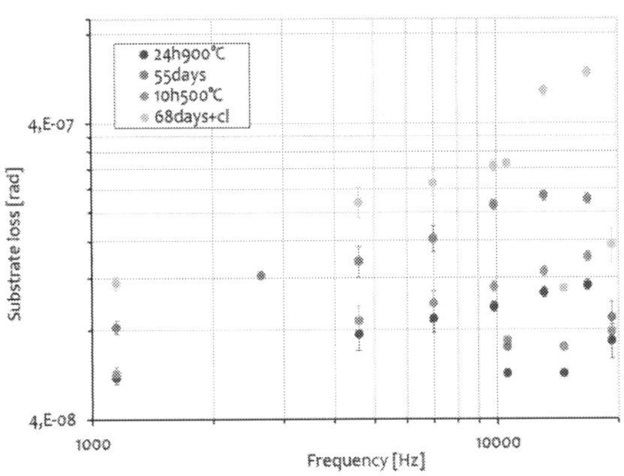

Figure 2. Loss angle dependence on frequency and mode families [18] for a SiO_2 sample produced by Corning of 3 inches diameter. Blue dots are the initial mechanical characterization, while red ones show the ageing effects (55 days later). Green dots are the characterization after an annealing treatment. In yellow the last characterization, that shows that the annealing does not stop ageing effects definitively.

In [17] a model that justify the existence of this family branching and predicting the level of losses is presented. The phenomenon could arise from excess losses inside the rough, not polished edge of the commercial samples and interests mainly the butterfly modes.
All these issues makes the characterization of the SiO_2 substrate not easy. In the following a strategy to solve these issues is shown.

5. Fused silica substrates preparation and characterization procedure

Due to the analytical method used to extract coating loss angle and also due to the issues of fused silica substrates, a precise procedure to prepare and characterize the samples is needed. It includes many mechanical characterizations through GeNS, and different preparation procedures regarding the SiO_2 substrate, like polishing of the lateral surface, annealing and vacuum storage. To reduce effects described in Section 4.2, a facility for the *CO_2 laser polishing* of the substrates barrel has been designed and realised in the Roma Tor Vergata University [19]. It includes:

- a CO_2 laser (maximum power $P_{max} = 15$ W),
- a red LED, for the alignment of the CO_2 beam,
- a lens (=150 mm), to adjust the CO_2 beam dimension,

- two galvo mirrors, to steer the CO_2 beam and perform many heating patterns,
- a rotary stage with a step motor, to rotate the samples during the polishing procedure.

In Fig. 3 it is possible to see the optical bench layout and some details of the components. With this facility it is possible to freely choose the set of parameters of the polishing process to optimize it, like rotation rate, number of rounds, heating pattern, beam power and beam size.

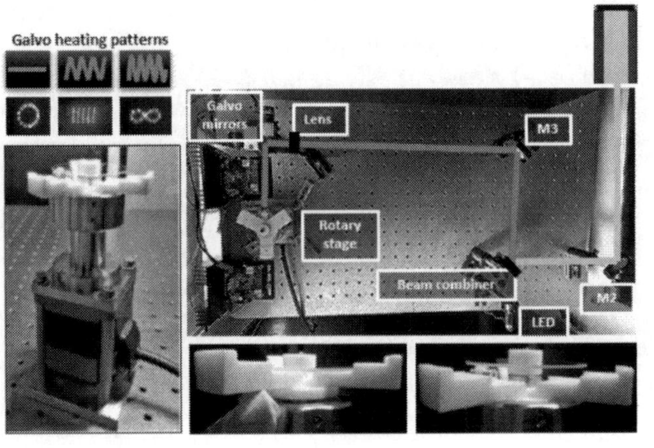

Figure 3. Optical bench layout of the CO_2 polishing facility in the University of Roma Tor Vergata, with details of part of the components, like galvo mirror heating patterns and rotary stage.

Other two procedures have been identified having effects on sample losses:

- *annealing* of the sample, consists of a thermal heating of the whole disk at high temperatures; for this procedure the sample has to be put inside a suitable container to protect it from any contamination that can arise during the heating; it allows to relax the structural frictions inside the samples;
- *vacuum storage* (still under study), consists of the storage of the samples in vacuum chamber for a suitable period; probably allows to de-absorb contaminants inside the unpolished (or not perfectly polished) lateral surface.

The effects of all the procedures can be quantified through mechanical characterization of the substrates. In Fig. 4 the mechanical characterizations of two different fused silica samples are shown: in each plot can be seen the CO_2 polishing effect on the sample mechanical losses (from blue initial characterization to green one). Particularly in Fig. 4(b) one can see that the CO_2 polishing of the barrel allows to maintain the sample losses almost stable in time.

6. Conclusions

Fused silica substrates can suffer loss angle variation over time due to different mechanisms, still under study. Their mechanical characterization is of fundamental importance to obtain a reliable estimation of the deposited coating loss angle.
A procedure to minimize or eliminate these loss angle variations and to mechanically characterize the substrates has been outlined. The substrate must undergo to a series of steps:

- initial mechanical characterization,
- CO_2 barrel polishing,
- annealing treatment and/or vacuum storage,
- second mechanical characterization.

This procedure has been already tested on some samples, showing good results in reducing both branching and ageing effects: looking at Fig. 4 the branching between butterfly and mixed

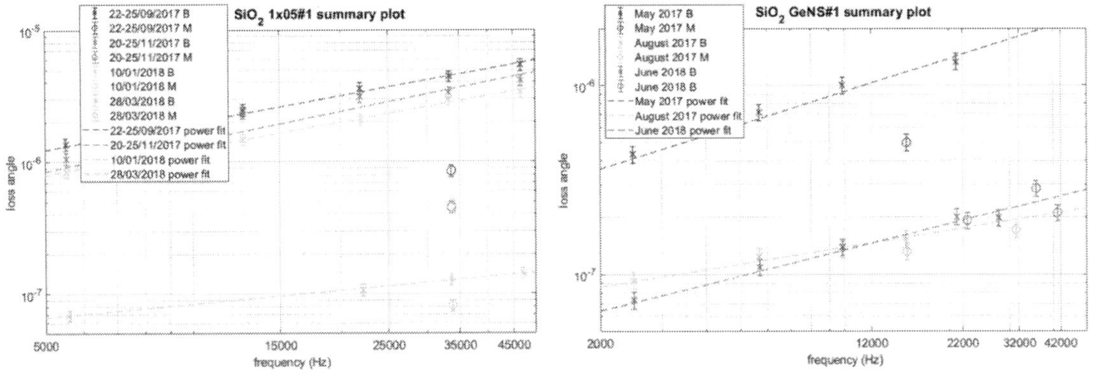

(a) (b)

Figure 4. (a) Mechanical characterization for a SiO_2 sample produced by Corning of 1 inch diameter. In blue, initial characterization; in pink effect of the annealing procedure; in light blue effect of the vacuum storage; in green effect of the polishing procedure. (b) Mechanical losses in a Suprasil Impex SiO_2 substrate of 2 inches diameter. In blue the initial characterization; in green the effect of the polishing procedure; in red the characterization made 10 month after the CO_2 polishing. In both plot, B stands for butterfly modes and M stands for mixed ones.

modes measured in different moments is reduced after the polishing procedure; furthermore, the measurements in Fig 4(b) shows that after one year mechanical losses remain constant, without ageing effects.
More investigation on annealing and vacuum storage is needed and further improvements of CO_2 polishing procedure are possible.

References

[1] Abbot B P et al. (LIGO Scientific Collaboration and Virgo Collaboration) 2016 *Phys. Rev. Lett.* **116** (061102)
[2] Abbot B P et al. (LIGO Scientific Collaboration and Virgo Collaboration) 2017 *Phys. Rev. Lett.* **119** (161101)
[3] Abbot B P et al. (LIGO Scientific Collaboration and Virgo Collaboration) 2017 *Phys. Rev. Lett.* **119** (141101)
[4] LIGO Scientific Collaboration, Virgo Collaboration, Fermi Gamma-Ray Burst Monitor and INTEGRAL 2017 *Astrophys. J.* **848** L13
[5] Penn S D et al. 2003 *Class. Quantum Grav.* **20** 2917
[6] Brown R 1828 *Philos. Mag.* **4** 161173
[7] A. Einstein 1905 *Ann. Phys. (Berl.)* **17** 549560
[8] Callen H B and Greene R F 1952 *Phys. Rev.* **86** 703
[9] Saulson P R 1990 *Phys. Rev. D* **42** 2437
[10] Harry G, Bodiya T P and DeSalvo R 2012 *Optical Coatings and Thermal Noise in Precision Measurement* Cambridge University Press New York
[11] Acernese F et al. (Virgo Collaboration) 2015 *Class. Quant. Grav.* **32** 024001
[12] Pinard L, Michel C, Sassolas B, Balzarini L, Degallaix J et al. 2017 *Appl. Opt.* **56** 4
[13] Granata M et al. 2016 *Phys. Rev. D* **93** 012007
[14] Flaminio R et al. 2010 *Class. Quantum Grav.* **27** 084030
[15] Martin I W et al 2009 *Class. Quantum Grav.* **26** 155012
[16] Cesarini E, Lorenzini M, Campagna E, Martelli F, Piergiovanni F, Vetrano F, Losurdo G, Cagnoli G 2009 *Rev. Sci. Instrum.* **80** (5) 053904
[17] Cagnoli G et al. 2017 *Phys. Lett. A* **382** 33
[18] Granata M, Amato A, Cagnoli G, Cassar Q, Cesarini E, Lorenzini M, Malhaire C, Montani M, Piergiovanni F VIR-0499A-16
[19] Lumaca D et al. VIR-0671A-17

IOP Conf. Series: Journal of Physics: Conf. Series **1226** (2019) 012024 doi:10.1088/1742-6596/1226/1/012024

LIDAL (Light Ion Detector for ALTEA): a compact Time-Of-Flight detector for radiation risk assessment in space

A Rizzo[1,2], C Berucci[1], C De Donato[2], L Di Fino[1], F R Lamastra[5,6], C La Tessa[3,4], G Masciantonio[2], R Messi[1,2], C Morone[1,2], P Picozza[1,2], M Rovituso[3], F Tommasino[3,4] and L Narici[1,2]

[1] University of Rome "Tor Vergata", Physics Department, via della Ricerca Scientifica 1, 00133 Rome, Italy
[2] INFN Sezione Roma "Tor Vergata", Via della Ricerca Scientifica 1, 00133 Rome, Italy
[3] Trento Institute for Fundamental Physics and Application - TIFPA INFN - Via Sommarive 14, 38123 Povo (Trento), Italy
[4] University of Trento, Physics Department, via Sommarive 14, 38123 Povo (Trento), Italy
[5] Enterprise Engineering Department, Via del Politecnico 1 00133 Rome, Italy
[6] INSTM - Research Unit Rome Tor Vergata

E-mail: alessandro.rizzo@roma2.infn.it

Abstract. Manned space missions towards Moon and Mars planned in the next decades require a reliable radiation risk assessment considering the long time exposure of astronauts (up to years) to different radiation fields. The radiation environment inside a human space habitat, generated by the interaction of the Galactic Cosmic Rays and occasionally of Solar Particle Events with the spacecraft hull, is peculiar due to its composition (ions from Hydrogen to Iron, knock out neutrons) and the large kinetic energy range of the particles. For this reason the risk assessment approach used for astronauts in space is quite different from the one used on Earth. In this approach the risk for astronauts is evaluated calculating factors which score the risk in function of physical characteristics of the single particle, like the *quality factor Q* (related to the radiation ionizing power) or the squared ratio between the charge (Z) and velocity (β) of the particle (Z^2/β^2).

LIDAL-ALTEA (Light Ion Detector for ALTEA - Anomalous Long Term Effects on Astronauts) is an experimental apparatus which will allow to evaluate for the first time in the field the Z^2/β^2 risk factor of the single detected particle on-board the International Space Station.

The LIDAL system is a Time-Of-Flight detector designed to work paired to three Silicon Detector Units of the ALTEA, which will measure the deposited energy of the passing particle. The velocity of the particle (β), calculated from the Time-Of-Flight measurement performed by LIDAL, allows to evaluate the particle electric charge once related to the deposited energy measured by ALTEA. A first LIDAL prototype has been developed by the University of Rome "Tor Vergata" and tested at TIFPA (Trento Insistute for Fundamental Physics Applications) proton beam line, in order to evaluate the timing performances of the detector. Results are briefly presented and the current status of the apparatus production is discussed in view of the launch scheduled for 2019.

1. Introduction

Next manned missions in space towards Moon and Mars require reliable radiation risk assessment for astronauts, considering their long time exposure (years) to radiation fields that occasionally can be quite high. For this reason a detailed characterization of the radiation spectra inside human space habitats is needed. LIDAL-ALTEA (Light Ion Detector for ALTEA - Anomalous Long Term Effects on Astronauts) is a compact apparatus currently developed to fly on-board the International Space Station (ISS) to characterize the radiation environment inside the human space habitat. In high latitude tracts of the ISS orbit, the radiation field inside ISS modules mimics the expected one in spacecrafts during deep space missions [1]. For this reason its detailed characterization might be crucial to plan manned missions far from Earth. LIDAL is a Time-Of-Flight (TOF) detector designed to work paired to three ALTEA Silicon Detector Units (SDUs), which measure the particle deposited energy in silicon. TOF measurement performed by LIDAL, used to evaluate the speed of the particle (β), allows to evaluate also its charge Z once related to the particle deposited energy measured by ALTEA (see Fig. 1). These quantities can be used to calculate the factor Z^2/β^2, which scores the exposure risk for astronauts [2].
Performances of the technological solutions chosen to reach the designed time resolution (less than 120 ps of σ), such as fast plastic scintillators, PMTs (Photo Multiplier Tubes), fast discriminator chip and High performance TDC (Time to Digital Converter) chip (HpTDC), have been verified by developing a first LIDAL prototype. The prototype has been tested at proton beam line at TIFPA (Trento Insistute for Fundamental Physics Applications) in Trento. The results, briefly presented in this work, allowed to design the final version of the detector. The Engineering Model of the apparatus will be ready by the end of October 2018, while the Flight Model is manifested to be sent on-board the ISS in spring 2019. Next sub-sections introduce the risk assessment in space and explain how the LIDAL-ALTEA apparatus fulfills technical requirements needed to have data for a reliable risk assessment for radiation exposure in space.

1.1. Radiation field in human space habitats

During manned missions in space the astronauts are exposed to the *so-called* primary radiation fields which surround the spacecraft and to the secondary radiation fields, generated by the interaction of the primaries with the spacecraft hull. The main components of the primary radiation fields are the Galactic Cosmic Rays (GCRs) and the Solar Particle Events (SPEs). GCRs consist of ions, from Hydrogen to Iron, with kinetic energies that peak around 2 GeV/A, where A is the mass number. This radiation field is isotropic and is always present in each point of the Galaxy, but, concerning the manned missions in Low Earth Orbit (like ISS), its dynamics depends upon the spacecraft position along its orbit. Particles from SPE (Solar Particle Events) are mostly protons emitted by flares or CME (Coronal Mass Ejections) of the Sun: therefore they can hit the spacecraft sporadically, depending also upon the position of the spacecraft.
The secondary radiation fields, generated by the interaction of primary particles with the spacecraft hull, consist in knock-out neutrons and ions generated by the nuclear fragmentation process. This component, summed up to the primary component, interacts with tissues and organs of astronauts.

1.2. Risk Assessment on Earth and in space: a different approach

The risk assessment for radiation exposure is based, both on Earth and in space, on two kinds of quantities: the physical quantities and the radio-protectional ones. Physical quantities are measured using a particle detector: the absorbed dose D is one of the most important ones and gives the infinitesimal deposited energy released in the infinitesimal volume of mass. The radio-protectional quantities are evaluated starting from the physical ones: the most important are the *Equivalent Dose* H_t, related to the *deterministic effects* and the *Effective Dose* E, related to *stochastic effects*. Deterministic effects appear after exposure to high doses: they are characterized by the existence of dose threshold above which, certainly, an effect on human tissue/organ is manifested. Stochastic effects can manifest after long exposure to quite low doses: there is no dose threshold and the linear correlation between the dose and the severity of effect is generally assumed but not demonstrated. The Equivalent (H_t) and Effective (E) doses are evaluated according to the equations:

$$\begin{aligned} H_t &= \sum_r w_r D_{r,t} \\ E &= \sum_t w_t H_t = \sum_{r,t} w_t w_r D_{r,t} \end{aligned} \tag{1}$$

where $D_{r,t}$ is the absorbed dose by a tissue t delivered by a radiation r, w_r and w_t are risk factors whose meaning is explained below. The factor w_r weighs the risk associated to a certain type of radiation in function of its ionization power. According to the International Commission on Radiological Protection (ICRP) prescriptions [3], w_r for a nuclear fragment is 20 times the w_r for a photon, due to the higher ionization power of the ion. The factor w_t weighs the tissue or organ radio-sensibility: tissues/organs with higher cell reproduction rate are more radio-sensible than those with lower cell reproduction rate [4].

For high doses (deterministic effects) the risk is evaluated using data from Hiroshima/Nagasaky nuclear bomb explosions, criticality accidents in nuclear plants and cell cultures exposed to high doses for example at accelerators. Results coming from the latter approach are also extrapolated to lower doses to evaluate the risks for stochastic effects, studied also with the epidemiological approach (statistical). Risk assessment for astronauts and evaluation of radio-protectional quantities in space, such as the equivalent dose, require a quite different approach because of the fundamental differences between the two radiation environments

. To evaluate the equivalent dose in space, a blind use of w_r tabulated by ICRP for exposure to the cosmic ions spectrum ($w_r = 20$ for every ion), cannot allow to differentiate the contribution of different ions interacting with the tissues or organs. For this reason in space the equivalent dose is calculated by replacing the factor w_r in eq.1 with the quality factor Q (defined according to ICRP [3]) which weighs the dangerousness of a particle in function of its Relative Biological Effectiveness (RBE) [3] and its Linear Energy Transfer (LET).

Nowadays it is known that also the equivalent dose is not the best radio-protectional quantity to use for risk assessment in space: the concept of equivalent dose itself is based on homogeneous irradiation of the tissue or organ. The exposure to cosmic ions field of tissue/organs is generally not homogeneous and the related risk is a function of the track position with respect to the cell. For these reasons other factors which can weigh accordingly the risk are currently under investigation. According to Ref. [2] a suitable factor related to the risk is the ratio Z^2/β^2 of the Bethe-Bloch equation. This ratio is proportional to the ionizing power of the particle: higher ionization powers mean more kinetic energy transferred by the particle to the tissue, where it can damage cells and DNA. The functional behaviour of the ratio can be explained by the following considerations. At a first glance the interaction between an ion and the atomic electrons in the tissue hit by the particle occurs through the ionic electric field, which grows with the particle charge Z. At the same time an ion with lower velocity will stand more time in proximity of the atoms, ionizing them more effectively and transferring to them more energy with respect to a fast traveling ion. For these reasons a heavier ion with lower kinetic energy can deliver to the tissues higher doses with respect those delivered by a light fast traveling ion.

LIDAL-ALTEA apparatus is designed to measure the physical quantities to calculate this ratio for the first time in space particle by particle. TOF measurements performed by LIDAL allow to measure the β factor, while the ion charge Z is evaluated through PID (Particle Identification as shown in Fig.1), where the ion discrimination is performed by applying Bethe-Bloch equation.

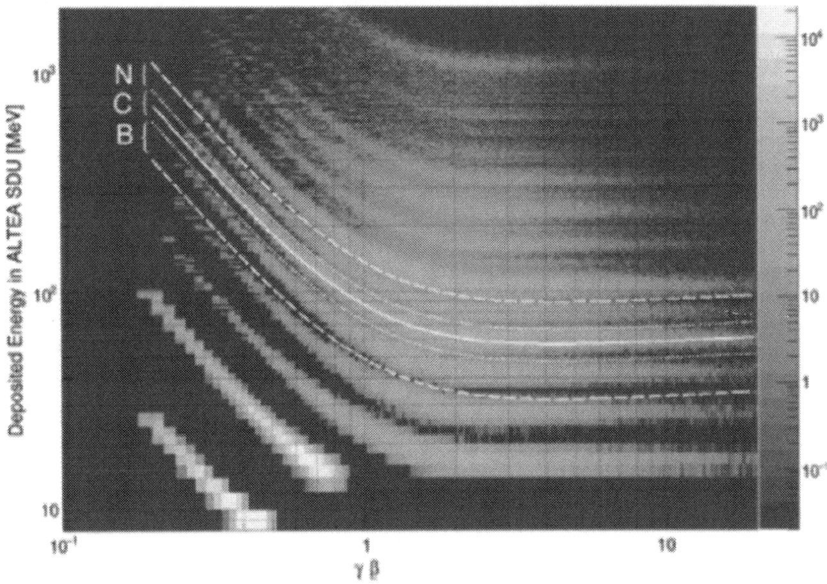

Figure 1. Expected spectrum acquired by LIDAL-ALTEA in 20 days of data taking. The white lines are the values of the Bethe Bloch equation for Boron, Carbon and Nitrogen ions. On Z axis (right) are shown the counts.

2. LIDAL detector and the prototype tests

In order to fulfill time resolution requirements for LIDAL apparatus (below 120 ps of σ, according to the simulations shown in Fig.1), state-of-the-art technological solutions suitable for space application have been chosen. Very fast plastic Scintillators (EJ-230 [5]) with 500 ps of rise time have been coupled, using PMMA light guides, to Hamamamatsu R988OU-110 PMTs (570 ps of rise time [6]). The output signal is processed by a NINO chip, a fast pre-amp and discriminator developed by CERN [7], and sent to a HpTDC chip used in High resolution mode (time resolution/channel 25 ps) [8]. A first prototype has been developed at "Tor Vergata" University and tested at TIFPA proton beam line to evaluate the timing performances of the detectors. The prototype uses two fast scintillators (dimensions $90\times25\times8$ mm) coupled, through two PMMA light guides placed at both ends of scintillator bar, to two PMTs. The distance between the scintillators can be set at 60 and 65 cm. More details about prototype apparatus and measurements results can be found in Ref.[9]. After equalization of the delay signal lines among the PMTs, several runs at different energies have been taken in the auto-trigger condition (AND condition between the two scintillators[1]). Two examples of TOF distributions are shown in Fig. 2 For protons in energy range between 70 and 230 MeV available at TIFPA beam line, time resolutions between 70 and 80 ps (σ) have been measured. The light collected at PMTs resulted to be plenty, giving PMT output signals larger than 2V in slightly undersupplied power conditions (about 900 V). This evidence led to a reduction of dimensions of final plastic scintillators to $80\times20\times4$ mm, a solution which can boost also the time performances in terms of signal timing development [9].

[1] No external trigger was applied in order to have conditions similar to the real ones in space.

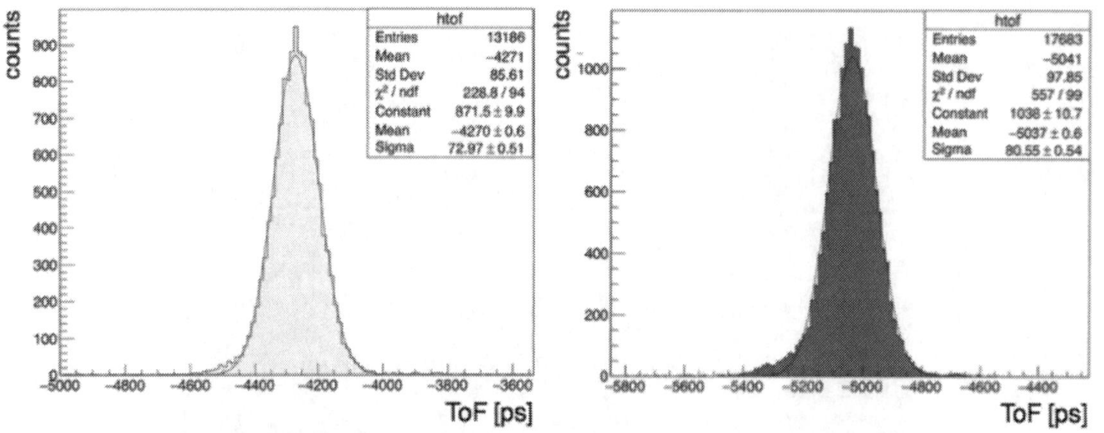

Figure 2. Left: TOF distribution for 159 MeV proton measured with a baseline of 65 cm. Time resolution (σ) is 73 ps. Right: TOF distribution for 159 MeV proton measured with a baseline of 60 cm. Time resolution (σ) is 81 ps. More details on the measurements can be found in Ref.[9].

3. Future perspective and conclusions

The LIDAL-ALTEA apparatus is manifested to be sent on-board the International Space Station in 2019. The results obtained during the measurements campaign at TIFPA proton beam line have been used to finalize the design of the detector. In particular the design of the electronic boards has been successfully completed and their production process started. The 32 scintillators units needed to realize the Engineering and the Flight Model of the apparatus are being realized at the University of "Tor Vergata". Each unit consists of a plastic scintillator coupled to two PMTs through two light guides. The optical contact between the scintillator and the light guides and between the light guides and PMTs is realized through optical cement EJ-500 [10]. Preliminary studies on tensile strength have been carried out on scintillators glued with light guides giving a breaking point higher than 7 MPa. The gluing procedure is done using specifically designed tools which allow to fulfill the required mechanical tolerance of 100 μm. Temperature and humidity are continuously checked using environment sensors. The PMTs are characterized before the gluing procedure by coupling them with scintillators realized for the first prototype and by exposing them to a Sr-90 source. After the gluing procedure the final scintillator unit is tested using the same radioactive source and the two responses are compared.
The LIDAL-ALTEA engineering model is scheduled to be ready in October 2018. Two beam tests are scheduled to characterize and calibrate the detector. The first will be held at TIFPA proton beam line in December and the second will be held at GSI (Helmholtzzentrum fur Schwerionenforschung GmbH) in February 2019, to calibrate the Flight Model apparatus with different species of ions.

References

[1] Narici L *et al.* 2017 *Scientific Reports* **7**, 1644.
[2] Cucinotta F 2013 *NASA TP* 217375 118121.
[3] *ICRP Publication* 92 2003 **4**.
[4] *ICRP Publication* 103 2007 **2-4**.
[5] *Fast timing EJ-228, EJ-230* available at: http://www.eljentechnology.com/products/plastic-scintillators/ej-228-ej-230. (Accessed July 2018).
[6] *Hamamatsu R9880U PMT series* available at: http://www.hamamatsu.com/resources/pdf/etd/R9880U_TPMH1321E.pdf (Accessed July 2018).

IOP Conf. Series: Journal of Physics: Conf. Series **1226** (2019) 012024 doi:10.1088/1742-6596/1226/1/012024

[7] Aghinolfi F *et al.* 2004 *IEEE Trans. Nucl. Sci.* **51** 1974-1978.
[8] Christiansen J 2004 *"HpTDC High Performance Time to Digital Converter Version 2.2"* available at https://cds.cern.ch/record/1067476/files/cer-002723234.pdf (accessed September 2018.
[9] Rizzo A *et al.* 2018 *Nuclear Inst. and Methods in Physics Research A* **898** 98104.
[10] *EJ-500 optical cement datasheet* available at : https://eljentechnology.com/images/products/data_sheets/*EJ*−500.*pdf* (Accessed July 2018).

IOP Conf. Series: Journal of Physics: Conf. Series **1226** (2019) 012025 doi:10.1088/1742-6596/1226/1/012025

Mathematical and geophysical methods for searching anomalies of the Radon signal related to earthquakes

F Ambrosino[1*], W De Cesare[2], V Roca[3] and C Sabbarese[1]

[1] Department of Mathematics and Physics, University of Campania "Luigi Vanvitelli", Viale Lincoln 5, 81100, Caserta, Italy
[2] Vesuvius Observatory - National Institute of Geophysics and Volcanology, Via Diocleziano 328, 80124, Napoli, Italy
[3] Department of Physics "Ettore Pancini", University of Napoli "Federico II", Via Cinthia 21, 80126, Napoli, Italy

[*]email: fabrizio.ambrosino@unicampania.it

Abstract. A detailed analysis of a time-series of the Radon (^{222}Rn) specific activity is carried out to identify anomalies that can be correlated with earthquakes occurrence in the monitored area. New hybrid methods are developed for this purpose and the advantages of each single component method are exploited. These methods are applied to two-years data series recorded continuously in the soils of a site within the seismo-volcanic area of Phlegrean Fields (Naples-Italy). Since the measurement system distinguishes and separately measures the ^{222}Rn and ^{220}Rn, an alternative estimation of the remote fraction of the gas and its anomalies has been also performed using the ^{220}Rn trend. The results of different methods are compared to recognize and to highlight Radon anomalies. Clear relationships have been found between anomalies and earthquakes of local origin and selected according to a specific formula of the earthquake preparation zone. The effectiveness of the methods and the goodness of the results are established by the high values of the cross correlation factors between the anomalies and the occurrence of the earthquakes.

1. Introduction

^{222}Rn is a radioactive, inert, water-soluble gas produced in Uranium decay series; its monitoring is very useful in relation to seismic phenomena thanks to the long half-life of *3.82 days*. In fact, the tectonic stress deformation developed inside the Earth's crust during seismic events causes fracturing of rock mass, develops the subsequent opening of various pathways and, consequently, anomalous surface emission of the gas can be produced. For instance, anomalies in the Radon signals can occur even at distances of a few hundred kilometres from a monitored earthquake, depending on its epicentre and magnitude [1-3]. Moreover, meteorological and environmental factors are important and not negligible because they influence the emanation and the exhalation rate of Radon in the superficial soil [4]. Thus, time-series analysis of the specific activity of Radon in soil gas can potentially reveal important information regarding crustal and surface processes, and also highlight precursor signals of the earthquake.

Research studies have been proposed in the framework of searching for correlation between Radon and earthquake occurrence, using different methodologies of time-series analysis [5-8]. This study addresses the application of four new methodologies with the aim of identifying anomalies instead of anomalous peaks, to better taking into account the several variation dependences of the signal. Three of the used methods are based on hybrid mathematical approaches, consisting of combinations of different algorithms, which provide greater precision by exploiting the advantages of each individual method [9-11]. The fourth method considers the trend of the ^{220}Rn isotope, which has a half-life of *55.6 seconds*, in order to eliminate the locally produced fraction of ^{222}Rn signal that is influenced by superficial and atmospheric effects [12].

2. Experimental setup

The investigated Radon time-series was recorded by continuous monitoring of the soil gas in a site inside an old dismissed railway gallery located within the Mount Olibano of the Phlegrean Fields area in Naples district (Italy). This area is interesting because characterized by high level of seismic risk caused by tectonics and volcano-tectonic activity [13]. ^{222}Rn and ^{220}Rn specific activities (expressed in Bq/m^3) were monitored by the RaMonA (Radon Monitoring and Acquisition) electrostatic collection device, by pumping the soils gas at a depth of *0.80 meters* into a chamber containing a Silicon detector for the α-spectrometric detection of the ionized Polonium isotopes, which are the progeny of the radioactive gas. Temperature, relative humidity, pressure parameters were also recorded by the device [14]. The analysed time-series was recorded in the period January 1st, 2016 – December 31st, 2017.

3. Radon time-series analysis

Three proposed methods are based on hybrid approaches, which have been implemented in Matlab® computing environment, and are the combination of:
1° Multiple Linear Regression and Auto-Regressive Moving Average statistical methods;
2° Empirical Mode Decomposition with Support Vector Regression techniques;
3° Singular Spectrum Analysis and a forecasting methodology.

The parametric combination 1° first creates a multiple linear regression model $y=b_0+b_ix_i+error$, performed by least-squares fit, linking Radon signal (y) and meteorological parameters (x_i) and, then, on the residue obtained from data minus the model, is performed the forecasting of the future values according to the linear and moving-average combination of previous data [15,16]. The adaptive combination 2° decomposes the signal (in trend, smooth, noise, periodic and seasonal components) and then the training of regression models on each component is made to predict the original series [12,17]. In detail, each component comes from the spline interpolation of the iterative minimum and maximum envelopes of the series with the local minima and maxima, and moreover each component is a function that has only one extreme between zero crossings, along with a mean value of zero. The parametric combination 3° consists first in the eigen-decomposition of the converted multi-dimensional matrix of the signal as eigenvalues and left and right orthonormal eigenvectors, where equal pairs of eigenvalues correspond to persistent oscillations; then the forecasting of the series is obtained by the sum of the last eigenvectors of the reconstructed signal along the diagonal of the multi-dimensional matrix [18]. These hybrid methods, essentially, analyse and estimate the different modulations in the time-series and predict the future signal in order to derive the anomalous signals. In addition, the three methodologies form effective strategies of time-series analysis combining the assets of each single method and producing a lower uncertainty compared with the individual methods more commonly used in the literature [19].

The fourth method 4° is named 'Remote Radon Estimation' (RRE) and is based on geophysical considerations; it makes possible to distinguish the remote component of the Radon signal, connected to the superficial phenomena, using the local variations of the ^{220}Rn isotope [12]. Since it has a half-life of *55.6 seconds* and, hence, a short migration length, the detected quantity is produced only locally and its variations are only due to local and surface effects. This methodology takes into account the specific activities, emanation coefficient, decay constant of the two isotopes, and the decay time of Radon within the tubes for pumping the gas inside the RaMonA chamber [12].

The identification of an anomaly is performed on the final signal obtained after the application of one of the four methods. Precisely, a positive or negative anomaly is located respectively above or below the 95% confidence intervals from the starting signal, where the two confidence bounds are calculated as confidence coefficient (*1.96*) multiplied by the standard deviation that is divided by the square root of the length of the signal. Finally, all the anomalies detected are normalized by subtracting to their value the mean value of the analysed series and this result is divided by the standard deviation.

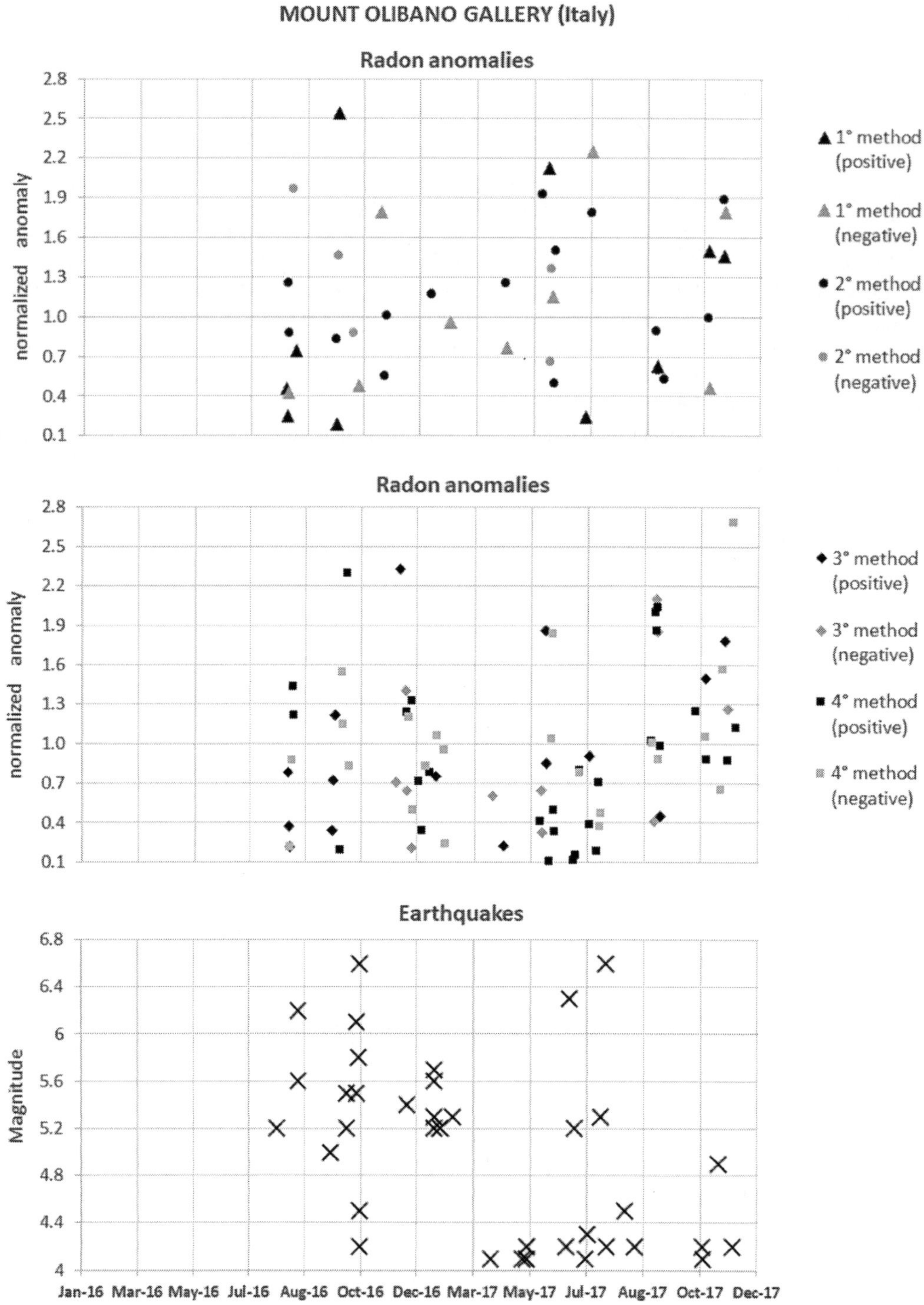

Figure 1. Anomalies (denoted as light symbols for negative and dark symbols for positive, in the first two graphs) found by the application of the four proposed methods for the analysis of the Radon time-series in Mount Olibano gallery, compared to 37 earthquakes selected to be of local origin and with the Dobrovolsky empirical formula (bottom graph).

4. Results

The results of the analysis of the climatic effects influencing the Radon signal show that the external temperature is the main driving force of the Radon.

Anomalies coming from the application of the four proposed methods on the monitored Radon time-series are displayed vs time in figure 1. In the same figure the earthquakes (taken from the *US Geological Survey* database - https://earthquake.usgs.gov) having influence on the Radon monitoring site of Mount Olibano gallery are also indicated. The earthquakes are selected of local origin and with the Dobrovolsky empirical formula [1, 20] that describes the area of tectonic deformation (with radius R expressed in *km*) caused by an earthquake with magnitude M: $R = 10^{0.43\,M}$. According to this formula, an earthquake that influenced the chosen Radon monitoring site is considered if the distance between the Mount Olibano and the earthquake location is within the radius of the manifestation zone of that earthquake. In figure 1, clear relationships are visible between the Radon anomalies provided by the four methods and the major earthquakes occurred in Italy and Greece. The strong correlations between the two events, persistent across the different approaches, show the effectiveness of this analysis and demonstrate the actual relationship between the anomalies of the Radon and the occurrence of earthquakes.

The overall results are supported and validated significantly by the cross-correlation statistical analysis:

- *85%* on average among the four methods;
- *87%* on average of values found for the four methods between Radon anomalies and earthquakes.

In particular, the RRE geophysical method shows cross-correlation statistical values of 80%, 86%, 84% and 88% with the results of the mathematical methods 1°, 2°, 3° and earthquakes, respectively.

5. Conclusions

This paper is focused on the development and application of four new methodologies used for the time-series analysis of the Radon signal recorded in a site of the interesting seismic area of Phlegrean Fields (Naples-Italy). The principal goal of this study is the anomalies identification, which are compared with earthquakes having influence in that area. The comparison of different mathematical and geophysical procedures is also another important goal. Significant results, supported by statistical evaluation, come from these analyses, which demonstrate the strength of the applied methods and prove as the Radon could be an important geochemical precursor of earthquakes occurred in the surroundings of the gas monitoring site. In particular, the RRE geophysical method based on ^{220}Rn local production and emission also provides results in good agreement with the others.

Acknowledgments

Authors wish to thank the staff of the Vesuvius Observatory, branch of the INGV (National Institute of Geophysics and Volcanology of Naples) for the Radon data management.

References

[1] Friedmann H 2012 *Radiat. Prot. Dosim.* **149 (2)** 177-84.
[2] İnan S and Seyis C 2010 *Acta Geophys.* **58 (5)** 828-37.
[3] Ghosh D, Deb A and Sengupta R 2009 *J. Appl. Geophys.* **69 (2)** 67-81.
[4] Yakovleva VS, Nagorsky PM, Kondratyeva AG and Mishina NV 2016 *IOP Conf. Ser.: Mater. Sci. Eng.* **142** 012051.
[5] Novikov A, Ulin S, Dmitrenko V, Vlasik K, Bychkova O, Petrenko D, Uteshev Z and Shustov A 2016 *J. Phys.: Conf. Ser.* **675** 042007.
[6] Dacey J 2010 *Phys. World* **23** 5.
[7] Ali Yalım H, Sandıkcıoğlu A, Ertuğrul O and Yıldız A 2012 *J. Environ Radioact.* **110** 7-12.
[8] Walia V, Virk HS and Bajwa SB 2006 *Pure Appl. Geophys.* **163 (4)** 711-21.
[9] Suhartono, Rahayu SP, Prastyo DD, Wijayanti DGP and Juliyanto 2017 *J. Phys.: Conf. Ser.* **890** 012160.

IOP Conf. Series: Journal of Physics: Conf. Series **1226** (2019) 012025 doi:10.1088/1742-6596/1226/1/012025

[10] Du P, Wang J, Yang W and Niu T 2018 *Renew. Energ.* **122** 533-50.
[11] Yahya NA, Samsudin R and Shabri A 2017 *J. Phys.: Conf. Ser.* **890** 012140.
[12] Sabbarese C, Ambrosino F, De Cicco F, Pugliese M, Quarto M and Roca V 2017 *Radiat. Prot. Dosim.* **177 (1-2)** 202-06.
[13] Piochi M, Bruno PP and De Astis G 2005 *Geochem. Gheophys.* **6 Q07005 (7)** 1-25.
[14] Sabbarese C, Ambrosino F, Buompane R, Pugliese M and Roca V 2017 *Appl. Radiat. Isot.* **122** 180-5.
[15] Razali SNAM, Rusiman MS, Zawawi NI and Arbin N 2018 *J. Phys.: Conf. Ser.* **995** 012041.
[16] De Cicco F, Pugliese M, Quarto M, Roca V, Sabbarese C, Savino F, Aquino I and De Cesare W 2017 *Environ. Earth Sci.* **76**, 317-27.
[17] Agustina SD, Mustakim, Okfalisa, Bella C and Ramadhan MA 2018 *J. Phys.: Conf. Ser.* **1028** 012240.
[18] Filho ASF and Lima GAR 2016 *J. Phys.: Conf. Ser.* **759** 012085.
[19] Wang J, Yang W, Du P and Li Y 2018 *Energy J.* **148** 59-78.
[20] Dobrovolsky IP, Zubkov SI and Miachkin VI 1979 *Pure Appl. Geophys.* **117 (5)** 1025-44.

IOP Conf. Series: Journal of Physics: Conf. Series **1226** (2019) 012026 doi:10.1088/1742-6596/1226/1/012026

Emergence of correlations in highly biased Consensus Models in seed initial configuration

M Di Vece[1] and F Corberi[1,2]

[1] Dipartimento di Fisica "E.R.Caianiello", Università di Salerno, via Giovanni Paolo II 132, 84084 Fisciano (SA), Italy

[2] INFN, Gruppo collegato di Salerno, and CNISM, Unità di Salerno, Università di Salerno, via Giovanni Paolo II 132, 84084 Fisciano (SA), Italy

E-mail: marryo36@hotmail.it

Abstract. We study the consensus probability in Voter Model and Invasion Process starting from a seed initial configuration. In the case where the opinions have the same strength or slightly different (weak bias) this function was computed analytically by Sood, Antal and Redner and depends only on the degree of the promoter individual. We check numerically through large scale simulations the above mentioned theory and we find that in the case of strong bias a correlation between the consensus probability and other centrality measures emerge and Sood et al's theory is broken.

1. Introduction

The manner in which opinions and ideas spread in a community is a topic of interest for the whole scientific community. From the beginning of the millennium scientists found that many real-world systems as the Worldwide Web or the Twitter Community can be well approximated by theoretical networks such as the Scale-Free network [1–3]. The latter one is a very heterogeneous network that well describes social media systems in which very few actors are very popular. In order to understand opinion spreading from a theoretical point of view, consensus models on such graphs can be considered. We will study toy models such as the Voter Model (VM) and the Invasion Process (IP)(see [4] for a review). In these models a binary variable (0 or 1) is assigned to each node representing its opinion state. The two models differ by their stochastic interaction rule as follows:

- in the VM a node i is chosen randomly with probability $1/N$ where N is the total number of nodes, then a second node j is picked among its neighbors, if j has a different opinion then i changes state.
- in the IP i is picked randomly in the network and then exports its state to one randomly chosen neighbor.

The dynamic process reaches an end when all nodes have the same opinion (*consensus state*). In regular graphs, where all nodes have the same number of connections k (also called degree), these two models are equivalent. In networks with an heterogeneous degree distribution this is no longer true and the two models have to be studied separately. If an opinion state is favoured with respect to the other a *bias parameter* s is introduced (for further details see [5]).

IOP Conf. Series: Journal of Physics: Conf. Series **1226** (2019) 012026 doi:10.1088/1742-6596/1226/1/012026

2. Sood et al's theory

In the following we will consider an uncorrelated Scale-Free Network built using the *uncorrelated configuration model* introduced by Catanzaro et al. [6]. We will use networks with size $N = 100$ and $N = 200$. Since we have checked that changing N gives similar results, we used $N = 100$ in cases when a larger statistical average was needed, and $N = 200$ to consider a larger range of degrees of the nodes. We will discuss only the most important results and leave the theorical details behind to the interested reader [4, 5].

2.1. Unbiased Case

Let us introduce the opinion state of node i, namely $\eta(i)$, and the degree-weighted moments ω_m

$$\omega_m = \frac{1}{N\langle k^m\rangle}\sum_i k_i^m \eta(i). \tag{1}$$

It can be shown [5, 7] that in heterogeneous networks ω_1 or ω_{-1} is conserved during the Voter Model process or the Invasion Process, respectively. From these conserved quantities it is easily recognized that the consensus 1 probability ε_1 for the two unbiased processes is

$$\varepsilon_1(\omega_1) = \omega_1 \tag{2}$$

for the VM, and

$$\varepsilon_1(\omega_{-1}) = \omega_{-1} \tag{3}$$

for the IP. Eqs. (2) and (3) are the main outcomes of the Sood et al's theory [5] for the unbiased case (i.e. $s = 0$). We will check numerically these results starting with a Seed Initial Configuration (SIC). In SIC there is only 1 node with opinion 1 at the start (seed or promoting individual) of the process, then

$$\varepsilon_1(k) = \frac{1}{N\langle k\rangle}k \tag{4}$$

for the VM, and

$$\varepsilon_1(k) = \frac{1}{N\langle k^{-1}\rangle}k^{-1} \tag{5}$$

for the IP.

Let us now discuss our numerical results. In Fig. 1 we plot $\varepsilon_1(k)$ for a scale-free network with $N = 100$. The relative error between theory and numerical data is 0.5% for the VM and 0.3% for the IP.

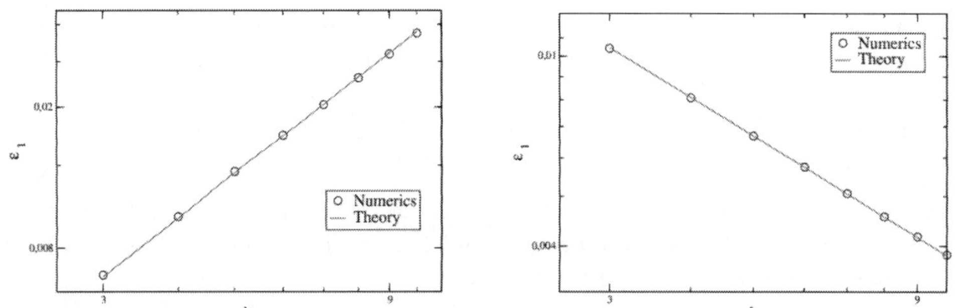

(a) VM (b) IP

Figure 1: ε_1 as a function of k for unbiased VM (left panel) and IP (right panel) on a scale-free network with $N = 100$. The numerical outcome is plotted with black circles and the theoretical curves of Eqs. (4) and (5) are plotted in red.

2.2. Biased Case

In the biased case (i.e. for $s > 0$) Sood et al. find the analytical form of ε_1 using the annealed approximation [5, 8] in the weak selection limit $s \ll 1$. They find a backward Kolmogorov equation for ε_1 of the form

$$\omega_1(1-\omega_1)\left[s\frac{\partial}{\partial\omega_1}+\frac{\langle k^2\rangle}{\langle k\rangle^2 N}\frac{\partial^2}{\partial\omega_1^2}\right]\varepsilon_1(\omega_1)=0 \tag{6}$$

for the biased VM, and

$$\frac{\omega_{-1}(1-\omega_{-1})}{\langle k\rangle\langle k^{-1}\rangle}\left[s\frac{\partial}{\partial\omega_{-1}}+\frac{1}{N}\frac{\partial^2}{\partial\omega_{-1}^2}\right]\varepsilon_1(\omega_{-1})=0 \tag{7}$$

for the biased IP.

These equations are very similar to those of a Random Walker with absorbing boundaries whose solution for the VM is

$$\varepsilon_1(\omega_1)=\frac{1-e^{-sN_{eff}\omega_1}}{1-e^{-sN_{eff}}} \tag{8}$$

where $N_{eff}=N\frac{\langle k\rangle^2}{\langle k^2\rangle}$, while for the IP it is

$$\varepsilon_1(\omega_{-1})=\frac{1-e^{-sN_{eff}\omega_{-1}}}{1-e^{-sN_{eff}}} \tag{9}$$

with $N_{eff}=N\langle k\rangle\langle k^{-1}\rangle$. In Fig. 2 we compare the theoretical functional form of ε_1 given by Eqs. (8) and (9) to the results of our simulations. The relative error between theory and numerical data is around $1-3\%$ both for the VM and for the IP, increasing with increasing s and k. We checked the nature of the discrepancies in increasing k varying the size of the network for fixed s and we found that for fixed k the error decreases increasing N, thus they are due to finite size effects. Let us note that in Fig. 2 we used different values of s for the two models. Sood et al. inferred that the weakly bias regime is given by $1/N_{eff} \ll s \ll 1$. For the analysed cases $(N_{eff})_{VM} < (N_{eff})_{IP}$ and so it is reasonable to use greater values of s for the VM.

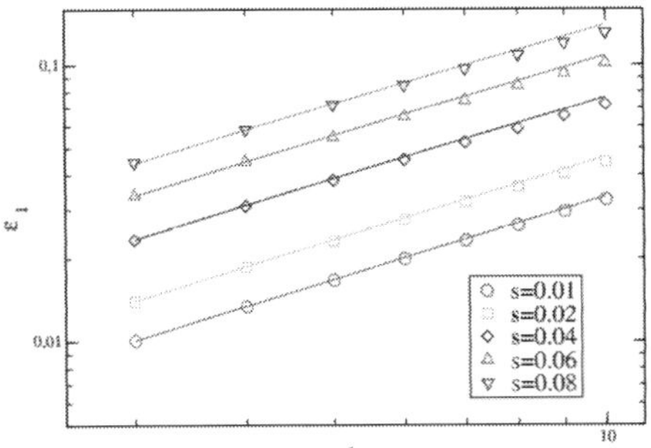

(a) VM

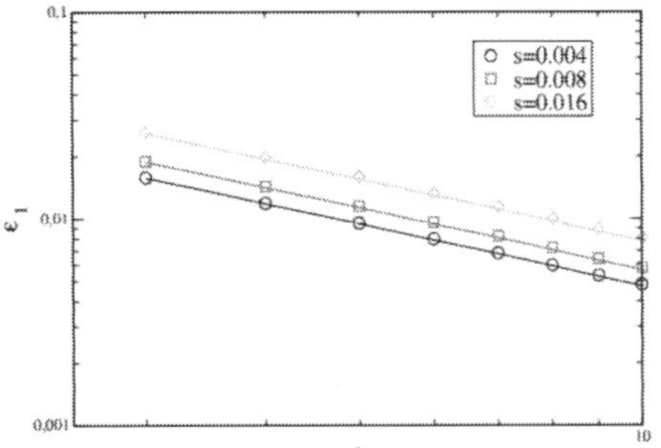

(b) IP

Figure 2: ε_1 as a function of k in the biased VM (left panel) and IP (right panel) for different values of the bias s with $N_d = 10^6$ stochastic realizations. Different colors correspond to different values of s. The process is simulated on a single realization of a scale-free with $N = 100$. The theoretical curves of Eqs. (8) and (9) are plotted with lines, the numerical outcomes are plotted with symbols.

IOP Conf. Series: Journal of Physics: Conf. Series **1226** (2019) 012026 doi:10.1088/1742-6596/1226/1/012026

2.3. *Strong bias*

We denote as *strong bias* the regime in which $s \gtrsim 1$. In this case we can see from Fig. 3 that Sood et al's theory (plotted with lines) is not in agreement with simulation data (symbols).

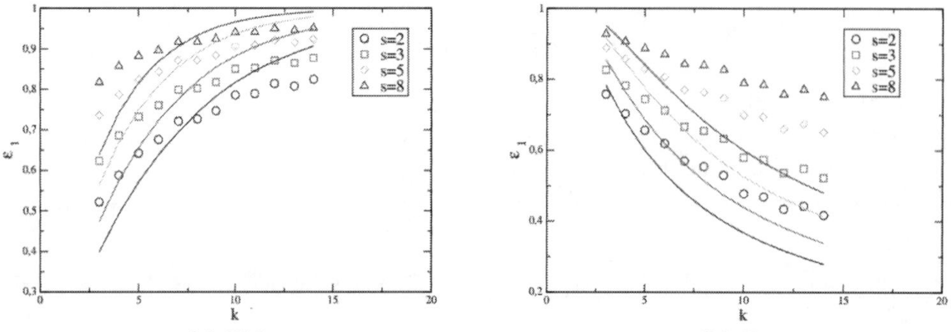

(a) VM **(b)** IP

Figure 3: ε_1 as a function of k for the biased VM (left panel) and IP (right panel) for different high values of the bias s, for $N_d = 10^6$ stochastic realizations. Different symbols correspond to different values of s. The process is simulated on a single realization of a scale-free network with $N = 200$. The numerical outcomes are plotted with circles and the Sood's curves of Eqs. (8) and (9) are plotted with lines.

Let us study the standard deviation σ_{ε_1} of ε_1. If ε_1 depends only on k then σ_{ε_1} should scale as $\frac{1}{\sqrt{N_d}}$, where N_d is the number of stochastic realizations used in the Montecarlo simulation, according to the central limit theorem. This is valid for $s \ll 1$ as we can see in Fig. 4a where σ_{ε_1} is well fitted by

$$\sigma_{\varepsilon_1} \simeq N_d^{-\beta} \tag{10}$$

with $\beta \approx 0.5$. However the central limit theorem is not obeyed for $s \gtrsim 1$ as we can see in Fig.4b. For IP we find similar results.

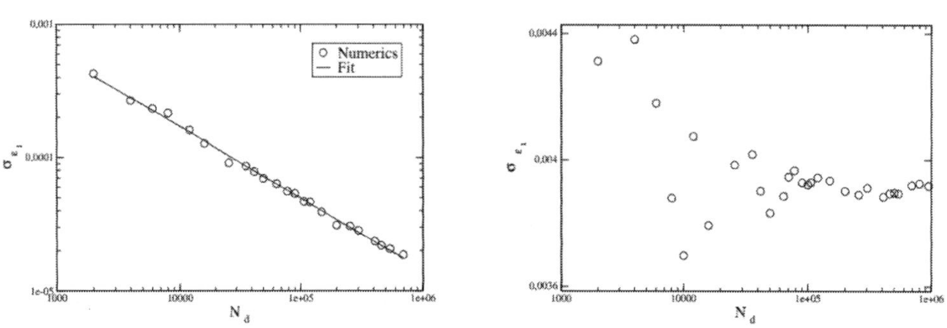

(a) VM weak bias **(b)** VM strong bias

Figure 4: σ_{ε_1} as a function of N_d for the nodes with $k = 3$ of the biased VM with $s = 0.01$ (left panel) and with $s = 3$ (right panel). The process is simulated on a single realization of a scale-free network with $N = 100$. The numerical outcome is plotted with black circles, the best fit of Eq. (10) is plotted with a black line.

IOP Conf. Series: Journal of Physics: Conf. Series **1226** (2019) 012026 doi:10.1088/1742-6596/1226/1/012026

3. Emergence of correlations between ε_1 and other topological measures

The fact that, for large s, ε_1 does not obey the central limit theorem suggests that it may depend on other quantities beside k. For this reason we study the possible correlations with other topological measures as the average neighbor's degree k_{nn} and the eigenvector centrality x (for further details on centrality measures see [9]). The eigenvector centrality has been used largely in the literature for quantifying the dynamic influence of a node [10].

Regarding k_{nn}, for the biased VM, when we increase s from the weak to the strong regime, we go from an uncorrelated behavior to an anticorrelated one for fixed k, as it can be seen in Fig. 5.

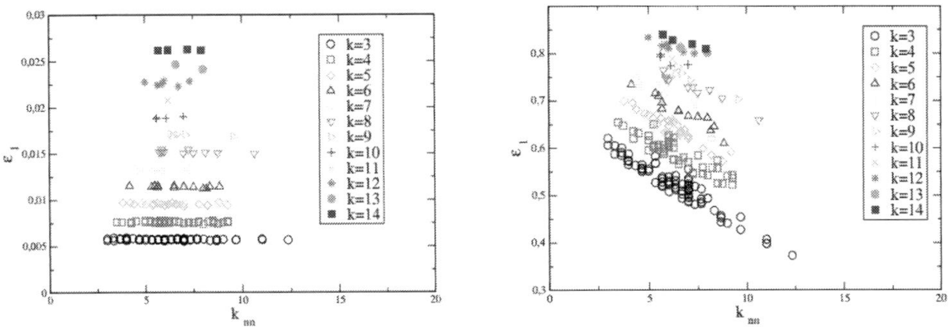

(a) VM weak s (b) VM strong s

Figure 5: ε_1 against k_{nn} for a biased VM with $s = 0.01$ (left panel) and $s = 3$ (right panel). Different symbols correspond to different values of k (see key). The process is simulated for $N_d = 10^6$ stochastic realizations of the dynamics on a single realization of a scale-free network with $N = 200$.

Instead, for the biased IP, we see in Fig. 6 that setting s in the strong bias regime induces correlations effects between ε_1 and k_{nn}.

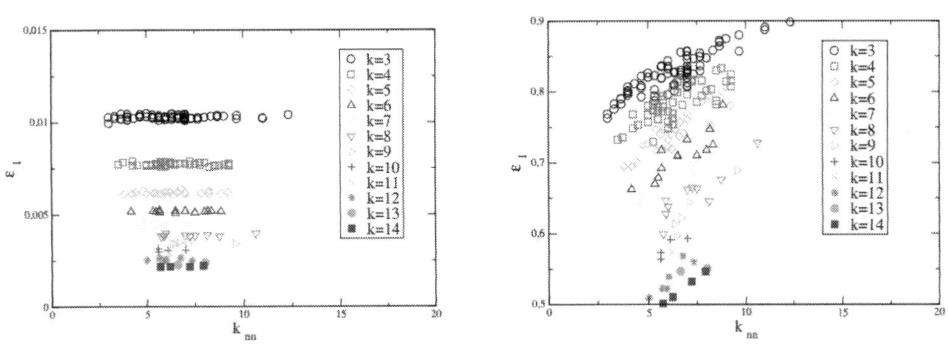

(a) IP weak s (b) IP strong s

Figure 6: ε_1 against k_{nn} for a biased IP with $s = 0.004$ (left panel) and $s = 3$ (right panel). Different symbols correspond to different values of k (see key). The process is simulated for $N_d = 10^6$ stochastic realizations of the dynamics on a single realization of a scale-free network with $N = 200$.

The correlations between ε_1 and the eigenvector centrality x are qualitatively similar to those with k_{nn}.

IOP Conf. Series: Journal of Physics: Conf. Series **1226** (2019) 012026 doi:10.1088/1742-6596/1226/1/012026

4. Conclusions

We studied the behavior of VM and IP with a null, weak and strong bias in seed initial configuration. While for null and weak bias the solutions given by Sood et al's theory are accurate, for strong bias they fail to predict the form of ε_1. We prove that this is due to the emergence of correlations between ε_1 and other topological measures such as k_{nn} and x in the biased VM or biased IP. Then we conclude that ε_1 does not depend only on k and a mathematical formulation that takes into account the presence of other correlations is needed.

Acknowledgments

We warmly thank Dr. C. Castellano for discussions.

References

[1] Barabàsi A L and Albert R 2002 *Rev. Mod. Phys.* **74** 47
[2] Albert R, Jeong H and Barabàsi A L 1999 *Nature* **401** 130âĂŞ131
[3] Aparicio S, Villazón-Terrazas J and Álvarez G 2015 *Entropy* **17** 5848-5867
[4] Castellano C, Fortunato S and Loreto V 2009 *Rev. Mod. Phys.* **81** 591
[5] Sood V, Andal T and Redner S 2008 *Phys. Rev. E.* **77** 041121
[6] Catanzaro M, Boguñá M and Pastor-Satorras R 2004 *Phys. Rev. E.***71** 027103
[7] Suchecki K, Eguiluz V M and San Miguel M 2005 *Europhys. Lett.* **69** (2) 228-234
[8] Pastor-Satorras R, Castellano C, Mieghem P V, Vespignani A 2006 *Physica A* **363** 1
[9] Barabàsi A L 2016 *Network Science*, Cambridge University Press
[10] Klemm K, Serrano M A, Eguiluz V M and San Miguel M 2012 *Sci.Rep.* **2**: 292

IOP Conf. Series: Journal of Physics: Conf. Series **1226** (2019) 012027 doi:10.1088/1742-6596/1226/1/012027

Inverse β-decay: a twin-model with boson fields

M Blasone[1,2], G Lambiase[1,2], G G Luciano[1,2] and L Petruzziello[1,2]

[1]Dipartimento di Fisica, Università di Salerno, Via Giovanni Paolo II, 132 I-84084 Fisciano (SA), Italy.

[2]INFN, Sezione di Napoli, Gruppo collegato di Salerno, I-84084 Fisciano (SA), Italy.

E-mail: blasone@sa.infn.it, lambiase@sa.infn.it, gluciano@sa.infn.it, lpetruzziello@na.infn.it

Abstract. In line with a series of previous works, in this paper we discuss the inverse β-decay of accelerated protons $p \to n + e^+ + \nu_e$. To make the calculation of the transition rate easier to handle, we focus on an ideal "twin-process", assuming the counterparts of lepton fields to be scalars. The rate of $p \to n$ conversions is evaluated in both the laboratory and comoving frames, showing that the two results perfectly agree due to the Unruh effect. In spite of the minimal setting, we stress that this model is an attempt to wash technical difficulties out of the analysis of the inverse β-decay. The future goal is to investigate the concerns recently raised in literature about the compatibility of the two rates when *neutrino mixing* is taken into account.

1. Introduction

The analysis of the inverse β-decay [1, 2] has proved to be enlightening for solving a puzzling question: is the Unruh effect [3] really mandatory for the consistency of quantum field theory (QFT) in curved background? As it is reasonable to expect, the decay rate of an accelerated particle in the laboratory and comoving frames must coincide. This outcome is prompted by the requirement of general covariance for the underlying formalism, which is built so that scalar quantities, such as the mean proper lifetime (i.e. the inverse of the decay rate) are invariant, and thus equal in any reference frame. In Refs. [1, 2], it was shown that this equality does indeed hold only when considering the inertial vacuum to be a thermal bath for the accelerated observer due to the Unruh effect, solving in this way the aforementioned dilemma.

Just when the issue of the inverse β-decay appeared to have been definitively settled, it hit again the headlines, becoming even more appealing. Indeed, if on the one hand in Refs. [1, 2] it was remarked that the decay properties of particles are not a frame-independent concept (according to what discussed in Ref. [4]), on the other hand the analysis there proposed completely overlooked *flavor mixing* of neutrinos, thus providing only a partial treatment of the problem.

The ingredient of mixing in the context of the accelerated proton decay was firstly introduced in Ref. [5] and later discussed in Refs. [6, 7, 8]. In Ref. [5], the authors highlighted a disagreement between the rates in the inertial and comoving frames, concluding that the contradiction had to be solved experimentally. On the other side, in Refs. [6, 7], such a paradox was overcome on purely theoretical basis, although conflicting approaches were taken into account. In Ref. [8], however, some criticism was exposed on the statements of Ref. [7], which was shown to fail to pinpoint the peculiar aspects of mixing in the problem under consideration.

IOP Conf. Series: Journal of Physics: Conf. Series **1226** (2019) 012027 doi:10.1088/1742-6596/1226/1/012027

Notwithstanding the clarification of the theoretical dichotomy, it is evident that the analysis of Ref. [6] needs further improvement, since it applies only within a particular approximation. So far, however, the non-triviality of calculations has prevented any development in this direction.

Aware of these technical difficulties, in the present paper we analyze a toy model[1] in which the inverse β-decay is addressed treating the electron and neutrino as scalar fields. By use of this minimalistic approach, we evaluate the rate of $p \to n$ conversions in both the laboratory and comoving frames, showing that the agreement is not affected by our assumption at all. In spite of the simplified setting, it should be emphasized that this model must be considered as a first step toward a more subtle investigation of the decay of accelerated particles in the context of flavor mixing. In this way, we hope to lay down the foundations for definitively clarifying the ambiguities recently raised in literature [5, 7].

The paper is organized as follows: In Sec. II we review the standard quantization of the scalar field in Minkowski and Rindler spacetimes. In Sec. III we discuss an inverse β decay-like process, showing that the outcome mimics the realistic case of the decay of an accelerated proton. The last Section is devoted to conclusions and outlook. Throughout the paper, we use natural units $\hbar = c = 1$ and the metric with the conventional timelike signature.

2. Quantization of the complex scalar field in Minkowski and Rindler spacetimes

Before going into the details of calculations, it is worth reviewing the quantization of a complex scalar field $\hat{\phi}$ in Minkowski and Rindler spacetimes [9]. In $3+1$ dimensions, the field expansion for an inertial (Minkowski) observer reads

$$\hat{\phi}(x) = \int \frac{d^3k}{\sqrt{(2\pi)^3 2\omega_k}} \left[\hat{a}_k e^{-ik\cdot x} + \hat{b}^\dagger_k e^{ik\cdot x}\right], \tag{1}$$

where $\omega_k = \sqrt{|\mathbf{k}|^2 + m^2}$, m is the mass of the field and $\hat{a}_k$ ($\hat{b}_k$) are the annihilators for Minkowski particles (antiparticles) with momentum $\mathbf{k}$. The plane waves in Eq. (1) satisfy the KG equation

$$\left(\Box + m^2\right)\phi(x) = 0\,. \tag{2}$$

In order to analyze the quantization procedure in Rindler spacetime, it is useful to express the coordinates appropriate to a uniformly accelerated (Rindler) observer in terms of the Minkowski ones. Choosing to leave the coordinates $\mathbf{x}_\perp \equiv (x, y)$ unchanged, namely, allowing the acceleration to have only the z component, we have

$$t = u \sinh v, \quad z = u \cosh v, \quad \mathbf{x}_\perp = \mathbf{x}'_\perp\,, \tag{3}$$

where $v \in (-\infty, \infty)$ is the time of the Rindler observer and $u \in (0, \infty)$, $\mathbf{x}'_\perp \in (-\infty, \infty)$ its spatial coordinates. In this way, the Minkowski metric becomes

$$g_{\mu\nu} = \mathrm{diag}\left(u^2, -1, -1, -1\right)\,. \tag{4}$$

We stress at this point that the world line of a Rindler observer with proper acceleration a is obtained by requiring $u = const = a^{-1}$ [10]. The field expansion in the right Rindler wedge ($z > |t|$) takes the form [1, 2, 10]

$$\hat{\phi}(x) = \int_0^\infty d\omega \int d^2k \left[\hat{\alpha}_{\mathrm{w}}\, \phi^{(\omega)}_{\mathrm{w}}(x) + \hat{\beta}^\dagger_{\mathrm{w}}\, \phi^{(-\omega)}_{\mathrm{w}}(x)\right], \tag{5}$$

[1] We refer to the simplified process described by this toy model as "twin process" for the standard inverse $\beta-$decay.

where $\mathrm{w} \equiv (\omega, \mathbf{k}_\perp)$ and $\omega \in (0, \infty)$ is the Rindler proper frequency. Here $\hat{\alpha}_\mathrm{w}$ ($\hat{\beta}_\mathrm{w}$) are the annihilators for Rindler particles (antiparticles) with frequency ω and transversal momentum $\mathbf{k}_\perp \equiv (k_x, k_y)$. The modes $\phi^{(\pm\omega)}_\mathrm{w}(x)$ are solutions of the KG equation in Rindler coordinates [10]:

$$\phi^{(\omega)}_\mathrm{w}(x) = \frac{1}{2\pi^2}\sqrt{\frac{\sinh\frac{\pi\omega}{a}}{a}}\, e^{-i\frac{\omega}{a}+i\mathbf{k}_\perp\cdot\mathbf{x}_\perp}\, K_{\frac{i\omega}{a}}(l\,u)\,, \tag{6}$$

where $l = \sqrt{|\mathbf{k}_\perp|^2 + m^2}$ and $K_{i\omega/a}(l\,u)$ is the modified Bessel function of the second kind. They are normalized with respect to the scalar product in Rindler coordinates [10, 11].

3. Inverse β-decay: an ideal "twin-model"

Let us now apply the above formalism to the calculation of the inverse β-decay rate. It is known that, according to the Standard Model, protons are stable. However, this is true only in the inertial case: for non-inertial protons, the accelerating field provides the rest energy difference between initial and final states, giving rise to the following process in the laboratory frame:

$$p \to n + e^+ + \nu_e\,. \tag{7}$$

In the original treatment of this decay, a semiclassical approximation is adopted in which the proton and neutron are seen as the ground and excited states of the nucleon Hamiltonian [1], while leptons are quantized in the usual way.

To make our analysis as transparent as possible, we can define in complete analogy with the above situation an ideal "twin-process", according to

$$g \to e + \bar{a} + b\,, \tag{8}$$

where the two-level system emulating the relation between p and n is now composed by two generic particles g and e, respectively[2]:

$$\hat{H}\,|g\rangle = m_g\,|g\rangle, \quad \hat{H}\,|e\rangle = m_e\,|e\rangle, \quad m_g < m_e\,, \tag{9}$$

and a, b are the counterparts of the electron and neutrino, which we assume for simplicity to be scalar fields. For small accelerations, one can show that the interaction action for the process Eq. (8) is [11]

$$\hat{S}_I = \int d^4x\,\sqrt{-g}\,\hat{Q}\left(\hat{\phi}^\dagger_a\,\hat{\phi}_b + \hat{\phi}^\dagger_b\,\hat{\phi}_a\right), \tag{10}$$

where $\hat{Q}$ is the Hermitian monopole

$$\hat{Q} = e^{i\hat{H}\tau}\,\hat{q}\,e^{-i\hat{H}\tau}\,\delta(x)\,\delta(y)\,\delta\big(u - a^{-1}\big)\,, \tag{11}$$

and $\lambda = |\langle g|\hat{q}|e\rangle|$ the effective coupling constant. Notice that the δ-functions reflect the fact that the particle g has a non-vanishing acceleration only along the z-axis.

From the point of view of the uniformly accelerated frame, the situation is significantly different from the one described above, being in this case the particle g at rest. With reference to the inverse β-decay, it has been shown that the $p \to n$ conversion becomes allowed due to the interaction of the proton with a flux of Rindler electrons and neutrinos popping out from the Unruh thermal bath, according to the following three processes [1, 2]:

$$(i)\;\; p^+ + e^- \to n + \nu_e\,, \qquad (ii)\;\; p^+ + \overline{\nu}_e \to n + e^+\,, \qquad (iii)\;\; p^+ + e^- + \overline{\nu}_e \to n\,. \tag{12}$$

[2] We remark that the labels g and e stand for *ground* and *excited* states, respectively.

IOP Conf. Series: Journal of Physics: Conf. Series **1226** (2019) 012027 doi:10.1088/1742-6596/1226/1/012027

Similarly, in our model we can suppose that, owing to the Unruh effect, the decay $g \to e$ occurs via the three following channels:

$$(i)\quad g + a \to e + b\,, \qquad (ii)\quad g + \bar{b} \to e + \bar{a}\,, \qquad (iii)\quad g + a + \bar{b} \to e\,. \tag{13}$$

Thus, our concern is to check whether the inertial and comoving decay rates match under our assumptions, in order for General Covariance to be preserved.

3.1. Inertial frame calculation

Starting from the interaction action Eq. (10), the tree-level transition amplitude for the process Eq. (8) reads

$$\mathcal{A}_{in} = \langle e| \otimes \langle \bar{a}\, b|\, \hat{S}_I\, |0\rangle \otimes |g\rangle\,. \tag{14}$$

Using the definition Eq. (11) of the Hermitian monopole, we obtain

$$\mathcal{A}_{in} = \lambda \int dt\, dz\, e^{i\Delta m \tau}\, \delta\left(u - a^{-1}\right) \langle \bar{a}\, b|\, \hat{\phi}_b^\dagger\, \hat{\phi}_a\, |0\rangle\,, \tag{15}$$

where $\Delta m \equiv m_e - m_g$ and τ is the proper time. By means of the properties of the δ-function, Eq. (15) can be rewritten as

$$\mathcal{A}_{in} = \frac{\lambda}{2(2\pi)^3\sqrt{\omega_a \omega_b}} \int_{-\infty}^{+\infty} d\tau\, e^{i\left[\Delta m \tau + (\omega_a + \omega_b)\sinh(a\tau)/a - \left(k_a^z + k_b^z\right)\cosh(a\tau)/a\right]}\,, \tag{16}$$

where we have used the field expansion Eq. (1). The differential transition rate is now defined as

$$\frac{d^6\mathcal{P}_{in}}{d^3k_a d^3k_b} \equiv |\mathcal{A}_{in}|^2 = \frac{\lambda^2}{4(2\pi)^6\omega_a\omega_b} \int_{-\infty}^{+\infty} d\tau_1\, e^{i\left[\Delta m \tau_1 + (\omega_a + \omega_b)\sinh(a\tau_1)/a - \left(k_a^z + k_b^z\right)\cosh(a\tau_1)/a\right]}$$
$$\times \int_{-\infty}^{+\infty} d\tau_2\, e^{-i\left[\Delta m \tau_2 + (\omega_a + \omega_b)\sinh(a\tau_2)/a - \left(k_a^z + k_b^z\right)\cosh(a\tau_2)/a\right]}\,. \tag{17}$$

To solve the integrals over proper time, we perform the transformation $\tau_{1/2} = s \pm \xi/2$, so that Eq. (17) becomes

$$\frac{d^6\mathcal{P}_{in}}{d^3k_a d^3k_b} = \frac{\lambda^2}{2(2\pi)^6\omega_a\omega_b} \int_{-\infty}^{+\infty} ds \int_{-\infty}^{+\infty} d\xi\, e^{i\left\{\Delta m\xi + 2a^{-1}\sinh(a\xi/2)\left[(\omega_a+\omega_b)\cosh(as) - \left(k_a^z + k_b^z\right)\sinh(as)\right]\right\}}\,. \tag{18}$$

With the further boost-transformations along the z-axis, namely $\omega'_\alpha = \omega_\alpha \cosh(as) - k_\alpha^z \sinh(as)$ and $k'^z_\alpha = -\omega_\alpha \sinh(as) + k_\alpha^z \cosh(as)$, $\alpha = \{a, b\}$, the integral over s can be singled out, giving the total proper time T of the two-level system made up of g and e.

Relabeling $\omega'_\alpha \to \omega_\alpha$, $k'^z_\alpha \to k_\alpha^z$, from Eq. (18) we now obtain

$$\frac{1}{T}\frac{d^6\mathcal{P}_{in}}{d^3k_a d^3k_b} = \frac{\lambda^2}{2(2\pi)^6\omega_a\omega_b} \int_{-\infty}^{+\infty} d\xi\, e^{i\left[\Delta m\xi + 2a^{-1}(\omega_a + \omega_b)\sinh(a\xi/2)\right]}\,, \tag{19}$$

which can be solved in terms of the modified Bessel function of the second kind [12]. By taking into account the result of the above integral and the general definition of decay rate

$$\Gamma = \frac{\mathcal{P}}{T}\,, \tag{20}$$

we are finally able to write the following integral expression for the inertial decay rate

$$\Gamma_{in} = \frac{2\lambda^2 e^{-\frac{\pi\Delta m}{a}}}{a(2\pi)^6} \int \frac{d^3k_a}{\omega_a} \int \frac{d^3k_b}{\omega_b} K_{2i\Delta m/a}\left(2\frac{\omega_a + \omega_b}{a}\right)\,. \tag{21}$$

The plot of Γ_{in} versus the acceleration is shown in the figure below.

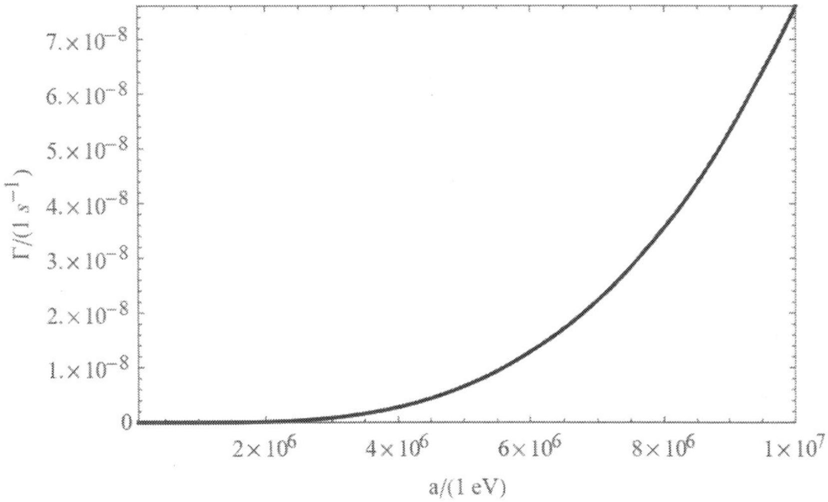

Figure 1. Plot of the decay rate as a function of the proper acceleration, for $m_a = 0.5\,\text{MeV}$, $m_b = 0.1\,\text{eV}$, $\Delta m = 1\,\text{MeV}$ and $\lambda = 10^{-12}\,\text{eV}^{-1}$. For $a \to 0$, one has $\Gamma \to 0$, which means that the particle g would be stable in absence of acceleration, as expected.

3.2. Comoving frame calculation

We now turn to the calculation of the decay rate in the comoving frame. For simplicity, we only deal with the process (i) in Eq. (13): similar considerations can be straightforwardly extended to the other two transitions.

In line with Eq. (14), the tree-level transition amplitude for the process (i) can be written as

$$\mathcal{A}_{(i)} \equiv \langle e| \otimes \langle b|\, \hat{S}_I\, |a\rangle \otimes |g\rangle = \frac{\lambda}{a} \int dv\, du\, e^{i\Delta m \tau}\, \delta\left(u - a^{-1}\right) \langle b|\, \hat{\phi}_b^{\dagger}\, \hat{\phi}_a\, |a\rangle\,. \tag{22}$$

Using the Rindler expansion Eq. (5) for the fields $\hat{\phi}_a$ and $\hat{\phi}_b$, we get

$$\mathcal{A}_{(i)} = \frac{\lambda \sqrt{\sinh\left(\frac{\pi\omega_a}{a}\right) \sinh\left(\frac{\pi\omega_b}{a}\right)}}{2\pi^3 a}\, \delta\left(\omega_a - \omega_b - \Delta m\right)\, K_{i\omega_a/a}\left(\frac{l_a}{a}\right) K_{i\omega_b/a}\left(\frac{l_b}{a}\right), \tag{23}$$

with l_a defined as in the previous section for each of the two fields.

There is now a subtle point to be considered: as mentioned above, in the comoving frame g interacts with a thermal bath of Rindler particles, giving rise to the processes Eq. (13). Hence, we must take account of this effect including the proper bosonic thermal factor of absorption/emission in the evaluation of the decay rate. For the process (i), we have

$$\frac{d^6\mathcal{P}_{(i)}}{d\omega_a d\omega_b d^2k_a d^2k_b} = |\mathcal{A}_{(i)}|^2\, n_B\left(\omega_a\right)\, \left[1 + n_B\left(\omega_b\right)\right], \tag{24}$$

where $n_B(\omega)$ is the Unruh thermal distribution for bosonic fields

$$n_B(\omega) = \frac{1}{e^{2\pi\omega/a} - 1}\,. \tag{25}$$

By inserting Eqs. (23) and (25) into Eq. (24) and recalling that the total proper lifetime of the two-level system is $T = 2\pi\delta(0)$, it follows that

$$\frac{1}{T}\frac{d^6\mathcal{P}_{(i)}}{d\omega_a d\omega_b d^2k_a d^2k_b} = \frac{\lambda^2\, e^{-\frac{\pi\Delta m}{a}}}{32\pi^7 a^2}\, \delta\left(\omega_a - \omega_b - \Delta m\right)\, K^2_{i\omega_a/a}\left(\frac{l_a}{a}\right) K^2_{i\omega_b/a}\left(\frac{l_b}{a}\right). \tag{26}$$

The decay rate can thus be obtained using the definition Eq. (20)

$$\Gamma_{(i)} = \frac{\lambda^2 e^{-\frac{\pi \Delta m}{a}}}{32\pi^7 a^2} \int_{\Delta m}^{+\infty} d\omega \int d^2k_a \int d^2k_b \, K^2_{i\omega/a}\left(\frac{l_a}{a}\right) K^2_{i(\omega-\Delta m)/a}\left(\frac{l_b}{a}\right). \tag{27}$$

If we now repeat similar calculations for the processes (ii) and (iii) and add up the three contributions, we finally get the following integral expression for the total (tree-level) decay rate:

$$\Gamma_{acc} \equiv \Gamma_{(i)} + \Gamma_{(ii)} + \Gamma_{(iii)} = \frac{\lambda^2 e^{-\frac{\pi \Delta m}{a}}}{32\pi^7 a^2} \int_{-\infty}^{+\infty} d\omega \int d^2k_a \int d^2k_b \, K^2_{i\omega/a}\left(\frac{l_a}{a}\right) K^2_{i(\omega-\Delta m)/a}\left(\frac{l_b}{a}\right). \tag{28}$$

The plot of Γ_{acc} as a function of the acceleration is identical to the one obtained in the inertial frame[3], as it can be seen from Fig. 1. Thus, the obtained result confirms the internal consistency of our theoretical model, in spite of its minimal setting.

4. Conclusions

In this paper, we have introduced a toy model for the investigation of the inverse β-decay of accelerated protons. To make the calculation of the transition rate as streamlined as possible, we have treated the electron and neutrino as scalar fields. Retracing the original steps of Ref. [1], we have shown that the General Covariance of QFT in curved background is preserved also within our simplified framework, provided that the Unruh effect is taken into account.

Clearly, this is just a preliminary study: as a next step, we plan to examine our formalism in the context of neutrino *flavor mixing*. Indeed, as discussed in Ref. [6], there is still an open debate on how to accommodate mixing effects in the analysis of the inverse β-decay without spoiling the agreement of the two rates, also in the light of a possible non-thermality of Unruh radiation arising in that case [13]. In particular, to clarify the ambiguities recently raised in literature [5, 7], the aim is to go beyond the approximation considered in Ref. [6], an effort that has not been successful so far due to the complexity of calculations. Further study in this direction is in progress [14].

Acknowledgements

We thank G. F. Aldi for technical discussion.

References

[1] Matsas G E A and Vanzella D A T 1999 *Phys. Rev.* D **59**, 094004; Matsas G E A and Vanzella D A T 2000 *Phys. Rev.* D **63**, 014010; Matsas G E A and Vanzella D A T 2001 *Phys. Rev. Lett.* **87**, 151301
[2] Suzuki H and Yamada K 2003 *Phys. Rev.* D **67**, 065002
[3] Unruh W G 1976 *Phys. Rev.* D **14**, 870
[4] Muller R 1997 *Phys. Rev.* D **56**, 953
[5] Ahluwalia D V, Labun L and Torrieri G 2016 *Eur. Phys. J.* A **52**, 189
[6] Blasone M, Lambiase G, Luciano G G and Petruzziello L 2018 *Phys. Rev.* D **97**, 105008
[7] Cozzella G, Fulling S A, Landulfo A G S, Matsas G E A and Vanzella D A T 2018 *Phys. Rev.* D **97**, 105022
[8] Blasone M, Lambiase G, Luciano G G and Petruzziello L 2018 *PoS CORFU* **2017**, 198
[9] Itzykson C and Zuber J B 1980 *Quantum Field Theory* (New York: McGrawHill)
[10] Takagi S 1986 *Prog. Theor. Phys. Suppl.* **88**, 1
[11] Birrell N D and Davies P C W 1982 *Quantum Fields in Curved Space* (Cambridge: Cambridge University Press)
[12] Gradshteyn I S and Ryzhik I M 1980 *Table of Integrals, Series and Products* (New York: Academic)
[13] Blasone M, Lambiase G and Luciano G G 2017 *Phys. Rev.* D **96**, 025023; Blasone M, Lambiase G and Luciano G G 2018 *J. Phys. Conf. Ser.* **956**, 012021
[14] Blasone M, Lambiase G, Luciano G G and Petruzziello L, *In preparation*

[3] Besides the numerical evaluation, it should be noted that the agreement between the two decay rates can also be proved analytically (see, for example, Ref. [2] for the case of the accelerated proton decay).